PRACTICAL SCIENCE

PRACTICAL SCIENCE

The role and reality of practical work
in school science

Edited by
Brian E. Woolnough

Open University Press
Milton Keynes · Philadelphia

Open University Press
Celtic Court
22 Ballmoor
Buckingham
MK18 1XW

and

1900 Frost Road, Suite 101
Bristol, PA 19007, USA

First Published 1991

British Library Cataloguing in Publication Data

Practical science.
   1. Great Britain. Secondary schools. Curriculum subjects:
Science. Practical work
I. Woolnough, Brian E. (Brian Ernest)
507.1241

   ISBN 0 335 09390 6
   ISBN 0 335 09389 2 (pbk)

Library of Congress Cataloging-in-Publication Data

Practical science: the role and reality of practical work in school
   science/edited by Brian E. Woolnough.
      p. cm.
   Includes bibliographical references.
   ISBN 0-335-09390-6. – ISBN 0-335-09389-2 (pbk.)
   1. Science–Study and teaching.   I. Woolnough, Brian E.
Q181.P63 1990
507.1-dc20   90-35681 CIP

Typeset by Vision Typesetting, Manchester
Printed in Great Britain by Biddles Limited, Guildford and King's Lynn

To Peggy,
to Anne and David,
to Ernest and Doris.

Contents

List of figures ix

List of tables x

Contributors xi

Preface xiv

Part I Introduction 1

  1 Setting the scene 3
    Brian E. Woolnough

Part II The role of practical science 11

  2 Practical work in school science: an analysis of current practice 13
    Pinchas Tamir

  3 The centrality of practical work in the Science/Technology/Society
    movement 21
    Robert E. Yager

  4 Practical science in low-income countries 31
    Terry Allsop

Part III The nature and purposes of practical science 41

  5 A means to an end: the role of processes in science education 43
    Robin Millar

  6 Practical work in science – a task-based approach? 53
    Richard Gott and Judith Mashiter

  7 Reconstructing theory from practical experience 67
    Richard F. Gunstone

  8 Episodes, and the purpose and conduct of practical work 78
    Richard T. White

Part IV The reality of practical science 87
9 Factors affecting success in science investigations 89
Kok-Aun Toh
10 School laboratory life 101
Joan Solomon
11 Gender differences in pupils' reactions to practical work 112
Patricia Murphy

Part V Complements to practical science 123
12 Simulation and laboratory practical activity 125
Vincent Lunetta and Avi Hofstein
13 Tackling technological tasks 138
Richard Kimbell

Part VI The assessment and evaluation of practical science 151
14 Principles of practical assessment 153
Bob Fairbrother
15 Assessment and evaluation in the science laboratory 167
Geoffrey J. Giddings, Avi Hofstein and Vincent Lunetta

Part VII Epilogue 179
16 Practical science as a holistic activity 181
Brian E. Woolnough

References 189
Index 201

List of figures

3.1   STS: domains for teaching and assessment   26
5.1   Sub-categories of 'practical skills'   51
6.1   Processes mediate procedural and conceptual understanding in the
      solution of a task   62
9.1   Performance in investigations (POSTTEST) by different treatment
      groups   96
9.2   Progression of performance over duration of study   97
9.3   Planning and performing scores   98
13.1 Traditional problem-solving model   141
13.2 An iterative active/reflective description of design activity   143
13.3 Concepts, procedures and manifestations: the axes of the
      assessment framework   145
13.4 Two levels within the assessment framework   146
14.1 An iterative approach to science as a problem-solving activity   159
16.1 Student attributes growing through practical science as a holistic
      activity   187

List of tables

2.1	Who does what in the science laboratory	16
2.2	Levels of inquiry in the science laboratory	16
3.1	Standard science and STS classrooms compared	23
3.2	Percentages of students who demonstrate ability to apply learning in science, by teaching method	27
3.3	Percentages of students with specific attitudes to science, by teaching method	28
3.4	Percentages of middle-school students who demonstrate their ability to perform in 14 science processes, by teaching method	29
8.1	A task of matching experiments with propositions	84
8.2	Example of a question testing linking between an experience and a proposition	85
9.1	Correlations between report sheets and supervisor's checklist	94
9.2	The four-group design	94
9.3	Intercorrelations among the predictor variables (ATTSCH, ATTSCI, ACADSC, PRKNOWL and SEX) and the criterion variable (POSTTEST) for groups 1 and 2	99
11.1	Some international survey results	114
12.1	Modes of simulation	127
12.2	Goals of laboratory activity	130
12.3	Phases of practical work and modes of simulation	131
12.4	Comparison of learning outcomes between simulation and practical work	132
15.1	Descriptive information for each scale in the Science Laboratory Environment Inventory	175

Contributors

Terry Allsop is a Lecturer in Science Education at the University of Oxford. After secondary school teaching, he worked in teacher education in Uganda and Hong Kong. More recently he has worked in China. Current interests include: science education in low-income countries; practical work; education in China; earth science in the science curriculum; and new models of initial teacher training. He is Course Tutor for the innovative 'internship' approach being pioneered in Oxford.

Bob Fairbrother is a Lecturer in Education at the Centre for Educational Studies, King's College, London. Formerly a school teacher, he has been examiner and chief examiner on subject committees of examining boards, on the GCSE Science Committee of the Secondary Examinations Council and on the Assessment Working Party of BTEC. He advised the Secondary Science Curriculum Review and the Graded Assessments in Science Project. He is currently involved in the development of Standard Assessment Tasks in Science for Key Stage 3 of the National Curriculum in England and Wales. He is Chairman of the Assessment and Examinations Committee of Association for Science Education (ASE) and has many publications on matters concerned with assessment.

Geoff Giddings is currently Associate Professor of Education and Head of the School of Curriculum Studies at Curtin University in Perth, Western Australia. He is also a project team member of the Australian Research Council-funded Key Centre of Research in Science and Mathematics Teaching, also located at Curtin University. He has taught extensively at both senior school and university level in Australia, North America and the United Kingdom. His current teaching and research interests include: innovation and change in the curriculum; curriculum development and leadership; and implementation and evaluation, particularly in the science education area.

Richard Gott is a Senior Lecturer in Education at the University of Durham. Following a period in comprehensive schools in the North of England, he joined the Assessment of Performance Unit in Leeds, where he was responsible for the development of the assessment of investigative work (with Patricia Murphy, now of the Open University) for pupils at ages 13 and 15. During his time at Durham he has become actively involved in in-service training work and in the development of Attainment Target 1, the exploration of science in the National Curriculum.

Richard Gunstone is an Associate Professor of Education at Monash University, Australia. Before joining Monash he taught physics, science and mathematics in Victoria high schools. His current research is concerned with the nature and meaning of constructivist views of learning, and the implications of these views for learning and teaching at both school and tertiary levels. In his teaching of pre- and in-service science teachers he attempts to respond to these implications.

Avi Hofstein is in the Department of Science Teaching, the Weizmann Institute, and is the Co-ordinating Inspector for Secondary School Chemistry in the Ministry of Education, Israel. He has provided leadership in science education since 1967.

Richard Kimbell is a Senior Lecturer in Education at Goldsmiths' College, University of London. He has taught technology in schools and been course director for undergraduate and postgraduate courses of teacher education. He is currently director of the Assessment of Performance Unit project in design and technology and co-director of the team developing Standard Assessment Tasks in technology for the National Curriculum.

Vincent Lunetta currently is Associate Dean for Research and Graduate Studies at Pennsylvania State University. He has served in the National Science Foundation and was a Professor at the University of Iowa. Some of his computer simulations in science and in education have been widely distributed. He has been the recipient of several awards for teaching and professional leadership in science education.

Judith Mashiter taught in a comprehensive school in the North-East of England before moving into the field of interactive video. Following research and development work on the use of new technology in science education at Newcastle University School of Education, she joined the National Interactive Video Centre.

Robin Millar is a Lecturer in Education at the University of York, where he is involved in the initial and in-service education of science teachers and in science curriculum development. He previously taught physics and science in several comprehensive schools. His research interests are science concept learning and conceptual change, and science studies in the context of science education. His publications include *Understanding Physics* (Unwin Hyman, 1989) and *Doing Science: Images of Science in Science Education* (Falmer Press, 1989).

Patricia Murphy is a Lecturer in Science and Technology at the School of Education of the Open University. Originally a chemistry teacher, she was a researcher and Deputy Director of the United Kingdom's Assessment of Performance Unit's science project at King's College, London. Her publications include research reports, reports for teachers, and articles on assessment and gender issues with particular respect to the science curriculum.

Joan Solomon is a Research Officer at Oxford University Department of Educational Studies. During thirty years' as a school teacher of physics, integrated science and technology, she initiated and wrote materials for the history and philosophy of science, *The Structure of Space* and *The Structure of Matter* (David & Charles, 1973), *Science in a Social Context* (Basil Blackwell, 1978), and *The Nature of Science* (Association for Science Education, 1989). Her research interests lie in children's conceptions and social issues; she founded Science Teachers In Research (STIR) (teacher research group). She teaches on the PGCE and in-service training courses at Oxford University.

Pinchas Tamir is Professor in the School of Education and the Israel Science Teaching Center, Hebrew University, Jerusalem and, since 1969, Director of the Israel Highschool Biology Project. His research interests include: curriculum development, implementation and evaluation; teacher education; science education with special emphasis on learning in the laboratory and on student assessment. He is author and editor of several books and of about two hundred research articles.

Kok-Aun Toh is Lecturer in Science Education at the Institute of Education (Singapore). He is Head of the Department of Science Education at the Institute, and is a past president of the Association for Science and Mathematics Education (Penang). His interests include science education, practical work and educational statistics.

Richard T. White came to Monash University, where he is now professor of Educational Psychology, after a decade as a high-school teacher of science. His interests are learning theory, science education, and research methods. His chief concern is to develop principles and practices that will promote a high quality of understanding, especially of science. His most recent work has been in association with an action research study, the Project for the Enhancement of Effective Learning.

Brian E. Woolnough is University Lecturer in Education (Physics) at Oxford University Department of Educational Studies. He is currently Honorary Editor of *Physics Education*, and his publications include *Practical Work in Science* (with Terry Allsop; Cambridge University Press, 1985) and *Physics Teaching in Schools, 1960–85* (Falmer Press, 1988). His many articles and reports focus on aspects of science and physics education, with particular reference to trends, practical work, and technology.

Robert Yager created one of the largest Science Education Centers in the United States, at the University of Iowa. He has worked with nearly 2,000 elementary and secondary teachers as they develop STS materials. He has served as president of numerous science education societies, including the National Science Teachers' Association (NSTA). He has directed numerous research projects, courses for graduate students, and teacher institutes.

Preface

Practical work plays a central and important part in the science programmes of schools in many countries. It often assumes the dominant, and dominating, role, involving teachers in a vast amount of time, effort and expense. And yet, it is possible for this work to be unfocused and unfulfilling, leading to much activity but little benefit. A recent talk by Professor Egglestone asked 'School Practical Science: It Is Practical, But Is It Science?'. We need to be clear about worthwhile aims for practical work, and to ask hard questions about its efficacy. I still remember, at the end of my first year as Head of Physics at my new school, sitting down and asking myself what I had been trying to do in the practical work during that year, what I had actually done, and whether that practice had been appropriate for those aims. My answers led to a resounding 'no' to the final question and a radical change and re-evaluation of my practical work in subsequent years. Much of my thinking and analysis since then has centred on these three questions. This book is an attempt to bring further light to them.

I have invited the leading authorities from around the world to share their latest thinking and research insights on various aspects of practical work in schools, in the hope that this will bring together and illuminate the important issues. I have been delighted that they have shown the commitment that they share with me to the importance of practical work in science by contributing so readily and lucidly to this volume. Though, deliberately, there is variation in the perceptions and approaches of the different authors, a common theme has recurred. Authors have accepted the importance of a holistic approach to practical work, moving away from the doing of standard practical exercises to verify some theory towards the tackling of investigational tasks to solve problems. It is this move, more than any other, which has focused and revitalised practical science in schools.

Part 1 introduces the book. In this Introduction I outline the agenda with some fundamental questions concerning practical work, which I then revisit in the Epilogue (Part VII), drawing together the key issues and setting down my personal view of the way forward. The other contributions are grouped in broad sections which, though overlapping as appropriate, enable the central issues to be addressed together.

Part II concentrates on the roles of practical science. Pinchas Tamir reviews current practice, while Bob Yager reaffirms the central role that practical work holds in the newer Science/Technology/Society-type courses, and Terry Allsop challenges us to reconsider the role and type of practical work appropriate for low-income countries.

Part III looks hard at the fundamental nature and purposes of practical work, and challenges some widely held assumptions. Robin Millar questions the validity of a process approach to doing science. Richard Gott and Judith Mashiter go further in their advocation of whole investigational tasks as being at the heart of doing science. Dick Gunstone and Richard White look at the ways in which practical work does, or does not, help theory and in which theory does, or does not, help practical.

In Part IV, three authors bring together research evidence into what actually is happening when students do practical science. Kok-Aun Toh reports his research findings into the factors that affect success in tackling a scientific investigation, Joan Solomon highlights the importance of the social interactions which occur, and Patricia Murphy reviews the different approaches that girls and boys bring to practical work.

It may seem strange to include chapters on the use of simulations, by Vincent Lunetta and Avi Hofstein, and on work in technology, by Richard Kimbell, in a book on practical work in science. These have been included as Part V, as they bring into fine relief aspects of science work. The simulation chapter sharpens those areas which are, and are not, best left to hands-on practical work. The parallels between problem-solving investigations in science and in technology are also important, stressing the way in which the one can illuminate and support the other. Scientists can learn much from the broader approaches in technology.

No book would be complete without chapters on assessment and evaluation (Part VI): these have been provided by Bob Fairbrother and by Geoffrey Giddings, Avi Hofstein and Vincent Lunetta.

I have learnt much from these contributions, which have influenced my own thinking significantly, and I wish to record my grateful thanks to all the contributors. I hope other readers will also find the collection stimulating.

Brian E. Woolnough

PART I

Introduction

CHAPTER 1

Setting the scene

Brian E. Woolnough

> . . . and with one loud *Worraworraworraworraworra* he jumped at the end of the tablecloth, pulled it to the ground, wrapped himself up in it three times, rolled to the other end of the room, and, after a terrible struggle, got his head into the daylight again, and said cheerfully: 'Have I won?'

This description of one of Tigger's adventures in *The House at Pooh Corner* is strongly suggestive of much pupil practical work in school science. There is much cheerful activity involved but, because the purpose of the exercise is not always clear, any sense of achievement or prizes won remains uncertain. In this book we hope to bring together the latest thinking of those around the world who have been researching and working in practical science in an endeavour to clarify some of the perplexing issues surrounding it.

A little background

Practical science, by which we mean the doing of experiments or practical exercises with scientific apparatus, usually in a science laboratory, has established a large and influential place in the science teaching of many countries. But even in the century or so of its existence (see Jenkins 1989; Layton 1973; Millar 1987), its nature and practice have changed considerably, especially over the last three decades. Since the late 1950s schools in the United Kingdom have moved from the standard 'cookbook' exercises to verify theory, through the 'guided discovery' of the Physical Science Study Committee (PSSC), Biological Sciences Curriculum Study (BSCS), Chem Study and Nuffield courses, to the various 'process' courses following *Science – A Process Approach* (American Association for the Advancement of Science 1967), such as *Warwick Process Science* (Screen 1986a), *Science in Process* (Wray *et al*. 1987) and the Oxfordshire Certificate of Educational Achievement (OCEA). These developed alongside the growth of an instrumental objectives approach to teaching and of investigational practical work encouraged by the Schools Council 5–13 project, Nuffield A-level physics, and the Assessment of Performance Unit

(APU) framework for science. Each of these initiatives was advocated on convincing grounds, in reaction to existing unsatisfactory practice, and yet each has been criticised in its turn.

It was not difficult to criticise the standard practical experiments designed to lead the pupil to the 'right result', even – or especially – when such were individualised [*sic*] through tightly prescribed worksheets. They did not lead to a good understanding of the underlying theory and they did not allow the pupils freedom to experience doing science for themselves. Similarly, the courses introduced in the 1960s, caricatured as guided discovery or 'stage-managed heurism', were criticised for failing to realise that the dual aims of being a scientist for a day and rediscovering predetermined scientific content were logically and practically incompatible (Stevens 1978; Wellington 1981). As courses sought to disentangle this dilemma by stressing the component processes of science, producing a spate of courses which focused on differing selections of underlying processes, skills, or process-skills, they in their turn drew criticism for their assertion that such processes existed in isolation, that they were distinctively scientific, that they were independent of theory and could be taught to pupils in a way which enabled them to be transferred to a broad range of scientific problems (Miller and Driver 1987; Jenkins 1987).

There is, of course, much variety of practice in schools. But despite the enormous amount of development, implementation and (a little) research, it is clear that science teachers and educators have failed to provide a rationale for, and style of, practical work for school science teaching that commands wide consensus in the profession. We hope that the reflections in this book will help further that goal.

Some outstanding questions

The aims of practical science

The first question that needs resolving relates to the purpose, or purposes, of doing practical work in science, and this in turn must be preceded by clarification of the purposes of teaching science as a whole. Should science be taught for cultural reasons, to enable pupils to appreciate both the discoveries and the ways of working in science, or for more vocational reasons, to provide pupils with such knowledge, skills and attitudes as they will find useful in later life? Are the same goals equally appropriate to all pupils? One of the dilemmas facing the 'science for all' movement is that society demands two things of school science – the provision of specialist manpower and of a scientifically literate citizenry. As Fensham (1985) commented, 'these two demands are conflicting and not complementary as was almost universally assumed in the first wave of the science curriculum movement'. Are the needs of technologically advanced countries the same as the needs of those with greater skill shortages? Though vocational training is clearly necessary in such countries, it is highly doubtful whether schools should be expected to provide such training. Conversely, it is not at all clear that the most helpful form of science for the developing countries is an importation of packaged courses developed in quite different cultures. Even the wisdom of teaching science in a science laboratory needs to be questioned in a country where the local environment might be the more appropriate base (Richmond 1978). With science courses being increasingly influenced by the Science/Technology/Society movement, what place does practical work have in them? Can the problem-solving approaches in practical science be related to the decision-making processes for developing citizenship?

Being good at doing science

Most rationales for science courses speak of pupils learning to do science as scientists do. But this assumes that we know how scientists work and that this is the most appropriate model for school science. Medawar (1969) is not alone in questioning the existence of a scientific method. He asks 'why are most scientists completely indifferent to – even contemptuous of – scientific methodology? . . . because what passes for scientific methodology is misrepresentation of what scientists do or ought to do'. There is no easy answer here. Perhaps we should be considering the various different ways in which different scientists work on different types of task – technicians requiring specific skills, pure scientists seeking 'the truth' and applied scientists solving problems.

If we accept that investigational practical work will be at the heart of pupils' laboratory work (investigations in which a scientific problem is set and the pupils are expected to plan their course of action, carry out the appropriate experiment, record and interpret the data and communicate the results, after the APU (1984) model of science as a problem-solving activity) we are still left with the question as to how pupils can best be helped to develop and improve. Do pupils learn by a 'step-up' approach whereby 'pupils are encouraged to master basic skills *first* and are *thereby* enabled to progress to more complex process skills and *eventually* practical investigations' (Bryce *et al.* 1987; emphasis added), or do they learn best by a holistic, experiential approach whereby they are encouraged to do small, but complete, investigations from the earliest stage, progressing to more difficult investigations later and picking up the appropriate skills when necessary? Do pupils develop their investigational ability best by learning the bits and putting them together or do they learn to do investigations by doing investigations? Such questions lead to the more pragmatic issue of the type of practical work we set our students. Do we set them 'standard exercises' to develop their skills? Do we give them 'routine experiments' to verify or discover some scientific theory? Do we provide them with 'hands-on experiences' to develop a feel for a particular phenomena? Or do we expect them to do 'practical investigations' to build up their own competence at working as problem-solving scientists?

The relationship between theory and practical work

One of the most persistent problems surrounding practical science concerns the relationship between theory and practice. Is the role of practical work to increase theoretical understanding? Is the role of theory to aid practical ability? Or should the two aspects of science be kept separate? Layton (1973) argues that the curriculum developers of the 1960s took too little notice of the problems involved in teaching these two aspects of science – its knowledge and its methodology – simultaneously. Fensham (1988) asserts that subsequent science courses have reduced 'the role of practical activity in science education to the enhancement of conceptual learning rather than being a source for learning essential skills and gaining confidence in applying scientific knowledge to solve real societal problems'. We have argued elsewhere (Woolnough and Allsop 1985) of the dangers of carelessly mixing the two aspects, or of making practical work subservient to theory, and suggested that we should initially separate these two aspects of science in our thinking, identify the quite separate justifications for developing both, and then reconsider their mutually supportive interaction. Others have gone further and asserted that the various skills and processes of science can be isolated and developed separately, sometimes even out of a scientific context, and then reassembled to produce the complete scientist. But it is clear that the processes of science are theory-laden, we observe not what is there but what our theoretical perceptions tell us is significant, and our success in applying those understandings is context-dependent. So can the two aspects of science be considered independently, should they always be considered as having an interdependent and interactive relationship? Or should we take a more holistic approach to the scientific approach to tackling scientific tasks, involving not only the cognitive and the psychomotor domains but also the affective?

Tacit knowledge

Perhaps the most tantalising problem relates to the extent to which our tacit knowledge, (or tacit knowing, as Polanyi would prefer), affects our ability to solve problems in practical science. I suspect it is considerable and yet, because of its very nature, we find it difficult to analyse. Polanyi said that 'we know more than we can tell', and this has been supported by the APU work which demonstrated that pupils were able to tackle practical investigations with considerable skill, even when they appeared to have little explicit mastery of the issues and processes involved; 'they were able to use their scientific knowledge in an investigation when they are unable to express the same knowledge in a written assessment' (Murphy 1985). Ravetz (1971) spoke of science as a craft activity, 'depending on a personal knowledge of particular things and a subtle judgement of their properties'. Often we try to reduce science to a series of skills and knowledge elements which we can measure, and ignore the all-important knowledge built up through years of personal experience which we cannot measure. Furthermore, we ignore the affective aspects of tacit knowing. How important are pupils' attitudes, motivation and belief in determining their success in a particular task? Probably of paramount importance, and yet often ignored. Bridgman defined scientific method as 'doing one's damndest . . . no holds barred'. Hodgkin (1985) asserts that 'believing is where learning starts. We know first, act on such knowledge and then get to know more . . . the activity of getting to know is compounded of feelings as well as of intellectual curiosity, of hunches as well as facts.' Hodson (1988) argues that we should 'prioritize the affective', and use practical experiments in science to build up pupils' confidence and self-esteem, for without that no success in science will be achieved. I have discussed this more fully elsewhere (Woolnough 1989a) and will return to it in the concluding remarks. I believe, with Fensham (1988), that this most understudied aspect of 'affect in action' is 'the major factor for improving science education'.

Transferability

Another question which we cannot ignore concerns the transferability of different attributes that pupils develop through doing practical science in schools. Transferability is usually claimed, or inferred, for a wide range of skills and processes acquired, but research evidence for such transferability is far from convincing. Direct transfer of a specific practical skill developed in the laboratory, such as the use of a microscope or the plotting of a graph, into a closely similar situation is non-contentious. But then the questions remain as to which skills are useful outside school, which tools and techniques are in general use, and whether schools are the most appropriate place in which to develop such skills. Whether the more general skills of planning, observing, interpreting, inferring, and so on, are transferable into quite different situations is far from proven. I suspect that the most important, possibly the only, attribute that a pupil acquires from practical science that is widely transferable is one of self-confidence. If the pupil has gained sufficient success and satisfaction from doing a wide range of investigations in school, that self-confidence will be transferable into a wider range of situations later. In Hodson's (1988) words:

> What may be transferable are certain attitudes and feelings of self worth . . . successful experience in one experiment may make children more determined and more interested in performing another experiment. The confidence arising from successfully designing an experiment might be a factor in helping children to stay at the task long enough to design a new experiment successfully.

And if that is true, the implications for the type of practical we set our pupils is clear but radical – they must provide success, stimulation and build self-confidence.

Progression

The next question is that of progression in practical work. What does a good investigation have which a less good one does not? How do pupils progress through their science education as their ability to do practical science increases? Such questions sound simple, but I suspect that there is no universal answer to them. Different pupils will develop at different rates and will react to different problems in quite different ways. I suspect that there is no common route along which all students travel, no set of meaningful milestones which can be prescribed and record the way each student progresses. Learning to be good at doing science follows more Bruner's roller-coaster model of learning than a uniform climb up a flight of stairs. Learning to be good at doing science is a much more unpredictable business. Scientists and science teachers can recognise a good and a less good investigation when they see them, but they would be rash if they tried to prescribe in advance how different investigations would develop. The 'quality' of an investigation is recognisable, but indefinable. The question of progression has taken on a particular urgency in England recently with the introduction of a national curriculum for science (Department of Education and Science (DES) 1989c). This national curriculum has been set into an assessment framework which necessitates ten levels of achievement for each aspect of the work, from the age of 5 to 16 (DES 1988a). One of the domains of science which is to be assessed concerns practical 'explorations', or investigations, and thus ten level statements must be prescribed and then assessed for pupils' investigational skills. The statements of levels of attainment have been varied in terms of context and complexity, and thus avoid any simplistic model of progression, but the problem remains for teacher assessment. Just how teachers will be able to judge to which level of attainment the pupil has progressed is not self-evident. How far progression in practical science may be prescribed in terms of specific criteria remains to be established.

Assessment

The final group of questions follow on from the issue of progression, and relate to methods of assessing practical skills. In particular, how can a valid form of assessment of practical work be established which encourages rather than inhibits the propagation of good pupil science activity? What is the effect of a particular assessment procedure on the way in which science is taught? If an assessment system seeks to measure specific practical skills, possibly even out of a scientific context, then emphasis and time will be given to developing those skills during teaching. If an assessment system demands that an extended practical investigation is to be done, then such will be included in the teaching programme. If the assessment system demands tight reliability of judgement, it may be forced to assess precisely trivial tasks which can be made teacher-proof. If it suggests a holistic form of assessment of the investigation as a whole against certain general criteria for good practice, it will have to rely on teachers' professional judgement and may lose some reliability for the sake of validity. The issues of reliability and validity, which ideally should jointly determine assessment practice, can in reality be contrary. I have argued elsewhere that an insistence on tight reliability against a closely prescribed set of specific criteria can reduce practical activity to a set of trivial exercises (Woolnough 1988b; 1989b). There has been much discussion about the problems of a norm-referenced assessment scheme and an advocation of criterion referencing in which assessment can be made against what a student 'knows, understands and can do'. But when we attempt to describe in advance a progressive series of steps along which all students should progress when tackling an open-ended scientific investigation, we run into fundamental problems concerning the very nature of real scientific activity – it does not progress along such tidy lines. Its very nature is unpredictable. We need to come to terms with such logical difficulties, without destroying the very nature of the practical science we are hoping to encourage.

So the agenda for our book is a formidable one. We do not pretend to be able to answer all of these questions. We hope, however, that we will engage with them constructively and thus further the teaching of practical science in which we all believe.

PART II

The role of practical science

Practical work in school science: an analysis of current practice

Pinchas Tamir

The uniqueness of practical work in school

Let us imagine a typical school day for most students. In most classes in most subjects the typical activities are talking, reading and writing. Suddenly, once, perhaps twice a week, there comes a change. The students enter the laboratory room where they will find taps, sinks, charts and pictures, various kinds of equipment, living organisms and an entirely different setting. Here, they are expected to do something with their hands, to observe, to measure, perhaps to smell, and from time to time even to plan, investigate and discover. Instead of sitting quietly they may talk freely with their team-mates, the atmosphere is much more relaxed and it is much easier to get the attention and the help of the teacher. In regular classes the atmosphere is often competitive, while in the laboratory, students usually co-operate and are expected to help each other.

In regular classes students' attention may easily be diverted from the learning task, while the concrete nature of laboratory work helps students focus their attention on the task at hand; even in case of diversion their attention can readily be regained. The laboratory offers many more opportunities for satisfying natural curiosity, for individual initiative, for independent work, for working in one's own time and for obtaining constant feedback regarding the effects of what one has been doing.

Practical work has gradually acquired an increasingly prominent place in the school science curriculum. One of the major changes advocated by the curriculum reform in the United States, the United Kingdom and elsewhere, is a new conception of the role of the school laboratory no longer as a merely illustrative and confirmatory adjunct to the learning of science concepts but, instead, as the centre of the instructional process.

· Two 'magic' key words have become associated with the 'new' school laboratories – 'discovery' and 'inquiry' – and people considered as the main conceptual leaders of the curriculum reform, such as Bruner, Gagné, Schwab, Piaget, Ausubel and Karplus, devoted much of their published work to issues related to learning in the laboratory. From their work, five major reasons may be offered as a rationale for the school science laboratory. First, science involves highly complex and abstract subject matter. Many students would fail to comprehend such concepts without the concrete props and opportunities for manipulation afforded in the laboratory (see, for example, Lawson 1975). Practical experiences are especially effective in inducing conceptual change (Lawson *et al.* 1989). Second, students' participation in actual investigations, employing and developing procedural knowledge often referred to as skills, is an essential component of learning science as inquiry (Schwab 1960; 1962). It gives students an opportunity to appreciate the spirit of science and promotes problem-solving, analytic, second generalising ability (Ausubel 1968). It allows the student to act like a real scientist (Bruner 1966), and develops important attitudes such as honesty, readiness to admit failure and critical assessment of results and of limitations, better known as scientific attitudes (see, for example, Henry 1975). Third, practical experiences, whether manipulative or intellectual, are qualitatively different from non-practical experiences and are essential for the development of skills and strategies with a wide range of generalisable effects (Gagné 1970).

> These skills function in concert in the mind of creative and critical thinker as he or she learns about the world . . . The skills are, in essence, learning tools essential for success and even for survival. Hence, if you help students improve their use of these creative and critical thinking skills you have helped them become more intelligent and helped them learn how to learn (Lawson *et al.* 1989, p. 31).

Fourth, the laboratory has been found to offer unique opportunities conductive to the identification, diagnosis and remediation of students' misconceptions (Driver and Bell 1985). Finally, students usually enjoy activities and practical work, and when they are offered and given a chance to experience meaningful and non-trivial experiences they become motivated and interested in science (Henry 1975; Lawson *et al.* 1989; Selmes *et al.* 1969).

The shift to laboratory-orientated courses in the 1960s was not based on 'hard' data showing the merits and superiority of such courses, but rather on the biases and convictions of the scientists and educational leaders of the curriculum reform. A comprehensive literature review up to the beginning of the 1980s indicated that research had still failed to support the effectiveness of the laboratory (Hofstein and Lunetta 1982). The reviewers suggested that the reason for this failure might be that 'past research study generally examined a relatively narrow range of teaching techniques, teacher and student characteristics and student outcomes' (p. 204). Another reason may be the ineffective use of practical work:

Science courses at all academic levels are . . . organized so that students waste many valuable hours in the laboratory collecting and manipulating empirical data which, at the very best, help them rediscover or exemplify principles that the instructor could present verbally and demonstrate visually in a matter of minutes (Ausubel 1968, p. 345).

Ausubel (1968, p. 345) asserts that

primary responsibility for transmitting the content of science should be delegated to teacher and textbook, whereas primary responsibility for transmitting appreciation of scientific method should be delegated to the laboratory. This does not imply that the laboratory and classroom should not be coordinated or that related substantive and methodological principles should not be considered together whenever relevant.

An even more extreme view of the desirable role of the laboratory is presented by Woolnough and Allsop (1985), who argue that 'the imposition of theory on practical work has had a detrimental effect on the development of scientific investigational skills' and advocate that we

stop using practical as a subservient strategy for teaching scientific concepts and knowledge since there are self-sufficient reasons for doing practical work in science and neither these, nor the aims concerning the teaching and understanding of scientific knowledge, are well served by the continual linking of practical work to the content syllabus of science (pp. 38–9).

If one accepts the need to offer a balanced curriculum in terms of content and process (Ausubel 1968), theory and practical work (Woolnough and Allsop 1985), and declarative and procedural knowledge (Lawson *et al.* 1989), the question remains about the optimal proportion of time that should be allocated to practical work.

How is the laboratory used?

A study of the literature enables a taxonomy of aims and objectives for practical work to be established. This can be structured under five main headings: understanding concepts (declarative knowledge); acquiring habits and capacities; gaining skills (procedural knowledge), including planning and design, performance, organisation, analysis and interpretation of data, and application to new situations; appreciating the nature of science; and developing attitudes. Such a taxonomy offers a broad list of potential outcomes. However, these can be achieved only if students are provided with the opportunity to be involved in the necessary experiences.

It might be enlightening to compare a typical laboratory lesson to a typical investigation carried out by a scientist. Table 2.1 compares a typical school

Table 2.1 Who does what in the science laboratory

Activity	Scientist's lab	School lab
Identifying problem for investigation	Scientist	Textbook or teacher
Formulating hypotheses	Scientist	Textbook or teacher
Designing procedures and experiments	Scientist	Textbook or teacher
Collecting data	Technician	Student
Drawing conclusions	Scientist	Students and teacher

Table 2.2 Levels of inquiry in the science laboratory

Level of inquiry	Problems	Procedures	Conclusions
Level 0	Given	Given	Given
Level 1	Given	Given	Open
Level 2	Given	Open	Open
Level 3	Open	Open	Open

laboratory with a typical scientist's laboratory in terms of who does what (following Anderson 1968). It will be seen that in a typical laboratory the student's work corresponds, by and large, to that of a technician.

Using the same categories as appear in Table 2.1, Pella (1961) identified different types of laboratory investigations. Following Herron (1971), these types may be arranged by the degree of openness and the demand for inquiry skills in four levels as shown in Table 2.2. In *level 0* problem, procedure and conclusions are all given and the only task remaining for the student is to collect data. In *level 1* problem and procedure are given and the student has to collect the data and draw the conclusions. In *level 2* only the problem is given and the student has to design the procedure, collect the data and draw conclusions. Finally, in *level 3*, the highest level of inquiry, the students have to do everything by themselves, beginning with problem formulation and ending with drawing conclusions.

As we have seen so far, the main criticism of practical work has been its undue emphasis on the lower levels, namely 0 and 1. However, even in classrooms which attempt to go beyond the two lower levels and increase the emphasis on inquiry, many students still perceive the laboratory as a place where they do things, but fail to see the connection between what they do and theory, and the place of the laboratory in the larger context of the scientific enterprise.

For many students inquiry-orientated laboratories present too many problems because of their demand for formal reasoning and the cognitive overload they often impose as a result of the need to apply simultaneously intellectual skills, practical skills and prior subject matter knowledge (Johnstone and Wham, 1982).

Tasker (1985) carried out extensive observations of the practical work of children aged 11–14. He identifies several reasons why children retain intuitive views in spite of science classroom experiences designed to teach the consensus science viewpoint. First, lessons are perceived by pupils as isolated events, not as parts of a related series of experiences as intended by the teacher. Second, the pupils' perceived purpose of the task is different from that of the teacher. Often teachers do not state the purpose. Even when they do they do not make sure that the pupils understand it. The tendency for pupils to construct as a purpose for a scientific classroom task either 'following the set instructions' or 'getting the right answer', was found in many classrooms which followed individualised programmes. Third, pupils fail to understand the relationship between the purpose of the investigation and the design of the experiment which they carry out. Fourth, pupils lack assumed prerequisite knowledge. Fifth, pupils are unable to grasp the 'mental set' required. Finally, pupils' perceptions relating to the significance of task outcomes achieved are not those assumed by the teacher.

The picture which emerges from the studies reported above is disappointing and discouraging, especially in light of the high expectations and far-reaching hopes for laboratory work. But there are more positive experiences which, even though relatively rare, still hold promise.

Lawson *et al.* (1989) provide convincing evidence of the effectiveness of an instructional approach designated as the learning cycle. This cycle involves three phases: exploring and identifying a pattern of regularity in the environment (exploration); discussing the pattern and introducing a term to refer to that pattern (term introduction, or invention); and applying the concept in new situations thereby discovering its deep and broad meaning (application, or discovery). According to Lawson *et al.* (1989, p. 76):

> The format of each phase of instruction is dictated by the role that the phase plays. The exploration phase is best suited to investigate nature and discover patterns of regularity. The laboratory format has been shown to be most effective in that role at least for highschool students . . . The learning cycle has many advantages over traditional instructional approaches especially when development of thinking skills is an important goal.

Using teacher interviews and questionnaires designed on the basis of these interviews, Dreyfus *et al.* (1982) found that inquiry-orientated laboratory activities had become an intergral and essential part of high school biology in Israel, well accepted and highly appreciated by the teachers and by the school administrators. In the judgement of the teachers the laboratories provide ample opportunities for students' investigations which involve organisms. Using structured observations, it was found that inquiry-orientated laboratories are significantly different from conventional ones. In inquiry-orientated laboratories the teachers are less direct; more planning takes place; processes of science receive more emphasis; there is more discussion after laboratory work has been carried out; and teachers give fewer instructions in front of the class; instead, they move around more, checking, probing, and supporting. Students are usually more active and they initiate ideas more frequently.

Brandwein *et al.* (1988) describe how they have been teaching successfully for many years the art of investigation to self-selected high school students. In their unique programme,

> as students go about their work they [are] encouraged to use a concept (brains on) to design an investigation (hands on) in planned laboratory work in search of meaning. Then brains on once again as students record their observations in the form of systematic assertion. To learn to originate . . . the young . . . have opportunity to innovate, to discover . . . to do an investigation or experiment (p. 277).

As a result of this unique laboratory-orientated approach,

> Forest Hills High School in New York, whose heterogeneous population embraced all students in the school district applying for enrollment, placed finalists and honorable mentions in the Science Talent Search in numbers comparable to those of the science high schools that selected students by requiring tests for entrance (p. 303).

An examination of the programmes described above indicates that when the practical work is properly employed, many of the expected outcomes are realised. The three key factors on which this realisation depends are the curriculum, the teacher and the prevailing procedures of students' assessment. Here we discuss the first two of these; assessment is discussed in Chapters 14 and 15.

The curriculum

The nature of classroom transactions is strongly dependent on the curriculum materials used, since these materials, in our case the laboratory manual, determine to a large extent the opportunities to learn offered to the students.

Tamir and Lunetta (1978) designed the Laboratory Analysis Inventory (LAI) which allows for a detailed analysis of the opportunities offered by a particular laboratory exercise in terms of the general organisation of work (for example, 'students work on different tasks and pool the results', or 'post-lab discussion required') as well as the inquiry skills called for (formulating a question, predicting results, hypothesising, observing, interpreting, explaining, applying). Using the LAI, Tamir and Lunetta (1981) found, for example, that the laboratory indeed plays a central role in the PSSC and Project Physics courses. Numerous experiences are provided in which students manipulate materials, gather data, make inferences and communicate the results in a variety of ways. At the same time, however, six important deficiencies were identified: First, there is no opportunity to identify probelms or to formulate hypotheses. Second, there are relatively few opportunities to design observation and measurement procedures. Third, there are even fewer opportunities for students to design experiments and to work according to their own design. Fourth, students are not encouraged sufficiently to discuss limitations and assumptions underlying their experiments. Fifth, students are not encouraged to share their efforts even in laboratory activities where that is appropriate. Finally, there are no provisions for post-laboratory discussion, consolidation of findings and analysis of their meaning. The results of such content analysis help us understand why certain lofty goals assigned to the laboratory have not, by and large, been achieved.

Content analyses of laboratory manuals reveal that even curricula claiming to be inquiry-orientated, such as the BSCS, PSSC and CHEM Study, offer, by and large, laboratory exercises which represent a very low level of inquiry (see Herron 1971; Tamir and Lunetta 1981). While some confirmatory laboratory exercises which aim at developing self-confidence as well as basic processes and techniques may be necessary, in our opinion, the majority should require students to engage in real problem-solving investigations, under different levels of guidance, according to particular goals and local conditions. A desirable way of facilitating inquiry is by assigning individual research projects which students do on their own under the guidance of the teacher.

While research on the nature of differences among laboratories in different disciplines is scarce, such differences obviously exist. In mathematics, for instance, the major purpose of concrete experiences appears to be the demonstration of relationships as well as assistance in problem-solving and intuitive rule-concept learning. In the physical sciences students make observations, measure and perform experiments. Yet, they often use instruments which translate the actual phenomena into data without being able to observe the actual phenomena directly. In a chemistry laboratory the students observe changes in colour or in appearance, hear explosions, notice different smells, and feel changes in temperature. Based on their perceptions, they have to infer what is happening. Atoms and molecules, electrovalent and covalent bonds can neither be seen nor touched; nevertheless, they constitute the conceptual basis for understanding what is happening.

In the physics laboratory students who work with electrical circuits are expected to explain their observations in terms of the behaviour of electrons which they are not able to see. This lack of direct perception is characteristic of much laboratory work in the physical sciences. While some investigations in biology and geology are similar to those in the physical sciences, biology and geology present more opportunities for direct observations of natural phenomena. Biology has the unique attribute of dealing with life, which implies care for plants and animals as well as employing special precautions and techniques in dealing with living materials. Emotions, such as compassion for animals or reluctance to perform dissections, are unique to laboratory work in biology. Another difference is that biology often requires some knowledge of physics and especially of chemistry. As a result of differences such as those described above, different guiding principles of inquiry have evolved for each discipline. Even within a discipline, such as biology, investigations in different subdisciplines such as anatomy, genetics and physiology may be guided by different principles, adapt different research strategies and employ different techniques.

The teacher

The teacher is undoubtedly the key factor in realising the potential of the laboratory. In order to be able to accomplish this mission, teachers need to be aware of the goals, potential, merits and difficulties of the school laboratory. Teaching in the laboratory requires a special approach to science (for example, science as inquiry), special instructional skills (running discussions before and after laboratory work), special management skills (budgeting time, managing small groups, guarding safety) and special attitudes (patience, tolerance of uncertainty, readiness to encounter failure, open-mindedness).

Careful preparation and planning on the part of the teacher, as well as assessment of performance and understanding of students through observations, diagnostic questioning, adequate homework and practical examinations are all essential. Lastly, the ability to integrate the laboratory with other instructional strategies and to motivate students will both make significant contributions to upgrading learning in the laboratory.

Unfortunately most teachers at present are ill prepared to teach effectively in the laboratory. A major reason is that 'most science teachers have themselves been brought up on a diet of content dominated cookery book-type practical work and many have got in the habit of propagating it themselves' (Woolnough and Allsop 1985, p. 80).

There is not doubt that teaching in the laboratory should receive much more attention in pre-service and in-service training than is currently given to it in most places.

With eyes to the future

The life of students and their learning in the laboratory can be improved by improving both the instructional materials and the instruction of the teachers themselves. In the words of Pickering (1980):

> The job of lab courses is to provide experience of doing science. While that potential is rarely achieved, the obstacles are organizational and not inherent in laboratory teaching itself . . . Massive amounts of money are not required to improve most programs. What is needed is more careful planning and precise thinking about educational objectives. By offering a genuine unvarnished scientific experience, a lab course can make a student into a better observer, a more careful and precise thinker and a more deliberative problem solver. And that is what education is all about.

CHAPTER 3

The centrality of practical work in the Science/Technology/Society movement

Robert E. Yager

For as long as there have been schools there have been attempts to make school science more meaningful. This chapter looks at the contribution in recent years of the Science/Technology/Society (STS) movement to these efforts.

One of the earliest assessments of school programmes was undertaken by Aristotle in 300 BC when he studied the schools of ancient Athens. He found two philosophies of education, one focused upon what society wanted a new generation to be like and a second focused upon skills students would find useful (practical) for living. He summarised his arguments as follows:

> There are doubts concerning the business of education since all people do not agree on those things which they would have a child taught, both with respect to improvement in virtue and a happy life; nor is it clear whether the object of it should be to improve the reason or rectify the morals. From the present mode of education we cannot determine with certainty to which men incline, whether to instruct a child in that which will be useful to him in life, or what tends to virtue, or what is excellent; for all these things have their separate defenders (quoted in Hurd, 1969).

The National Science Teachers' Association (NSTA) in the United States has endorsed a broad definition for STS. In its *Position Statement*, the NSTA (in press) proclaims:

> STS is the term applied to the latest effort to provide a real world context for the study of science and for the pursuit of science itself. It is a term that elevates science education rhetoric to a position beyond curriculum and the ensuing debate about the scope and sequence of basic concepts and process skills. STS includes the whole spectrum of critical incidents in the education process,

including goals, curriculum, instructional strategies, evaluation, and teacher preparation/performance. One cannot 'do' STS by adding certain topics and lessons to the curriculum, course outline, or textbook. Students must be involved with goal setting, with planning procedures, with locating information, and with evaluating them all. Basic to STS efforts is the production of an informed citizenry capable of making crucial decisions about current problems and taking personal actions as a result of these decisions. STS means focusing upon current issues and attempts at their resolution as the best way of preparing people for current and future citizenship roles. This means identifying local, regional, national, and international problems with students, planning for individual and group activities which address them, and moving to actions designed to resolve the issues investigated. Students are involved in the total process; they are not recipients of whatever a pre-determined curriculum or the teacher dictates. There are no concepts and/or processes unique to STS; instead STS provides a setting and a reason for considering basic science and technology concepts and processes. It means determining ways that these basic ideas and skills can be seen as useful. STS means focusing on real-world problems instead of starting with concepts and processes which teachers and curriculum developers argue in terms of usefulness to students.

If STS were to become the mechanism for changing typical school science, with its teacher/textbook focus on basic science concepts and processes, into a more practical study of student questions, the usefulness of science and technology concepts and processes could be discovered by students for themselves (rather than merely presented to them as desired outcomes proclaimed by teachers). Typical science classrooms can be contrasted with STS classrooms in several ways, as shown in Table 3.1.

Practical work is often defined as typical laboratory work where students encounter ideas and principles at first hand. To some it merely means 'hands-on' science. However, the typical science laboratory is not a laboratory at all if that term is reserved to describe a place where students can go to test out their own ideas and/or their own explanations for objects and events they have encountered as they have explored their own curiosities about the universe in which they have found themselves. When analyses have been made of US laboratory exercises that accompany textbook series (Fuhrman *et al.* 1982), it was found that in excess of 90 per cent of the so-called laboratories are verification activities, that no real investigations are conducted, that the correct answers to the activities are given before the laboratory work begins. This raises questions as to how 'practical' such work is. Brandwein (1981) has asserted that we would have a revolution in science education if every student were to have even one real experience with the whole science process (that is, identifying a problem, proposing possible explanations, and devising tests to determine the validity of a particular explanation).

Practical work in science implies that the activity represents science and the particular effort will be useful to the student. Some activities called 'practical' by

Table 3.1 Standard science and STS classrooms compared

Standard	STS
• Survey of major concepts found in standard textbooks	• Identification of problems with local interest/impact
• Use of laboratories and activities suggested in textbook and accompanying lab manual	• Use of local resources (human and material) to locate information that can be used in problem resolution
• Passive involvement of students assimilating information provided by teacher and textbook	• Active involvement of students in seeking useful information
• Learning being contained in the classroom for a series of periods over the school year	• Teaching going beyond a given series of class sessions, a given meeting room, or a given educational structure
• A focus on information proclaimed important for students to master	• A focus on personal impact, often starting with student curiosity and concerns
• A view that content is the information included and explained in textbooks and teacher lectures	• A view that content is *not* something that merely exists for student mastery simply because it is recorded in print
• Practice of basic process skills, but little attention to them in terms of evaluation	• Process skills played down
• No attention to career awareness other than an occasional reference to scientists and their discoveries (most of these long after their lifetimes)	• A focus on career awareness, especially careers that relate to science and technology and not merely those related to scientific research, medicine, and engineering
• Students concentrating on problems provided by teachers and textbooks	• Students performing in citizenship roles as they attempt to resolve issues they have identified.
• Learning occurring only in the classroom as a part of the school's science department	• Study being visible in a given institution and in a specific community
• Science being a study of information where teachers control how much students acquire	• Science being an experience students are encouraged to have
• Learning focusing on current explanations and understandings; little concern for the use of information beyond classroom and performance on texts	• Learning with a focus on the future and what it may be like

teachers, textbook authors, and curriculum developers are not practical (that is to say, useful and/or meaningful) for the students who experience them. Ideally, all practical work in science classrooms would exemplify the basic ingredients of science and would be practical for each student. In actuality it rarely is for even the most talented students. Instead, it is too often another kind of hurdle that students must master if teachers are to label them as exceptional students of textbook/school science. To be practical for the students, the work teachers require must be demonstrably useful in their lives. In an idealised STS situation, whether in a typical classroom/discussion format or in a so-called laboratory, the work is practical by definition because it focuses on the students' own questions, their explanations, their tests of validity for such explanations, and their own actions following problem resolution.

STS means a broader view of school science. However, it does not mean merely adding two components to science – technology and society. Science, technology and society can each be viewed as a continuum, with the concepts and processes understood and practised by professional scientists, technologists, and sociologists at one extreme, and the uninformed or uninitiated human who is trying to learn, that is trying to prepare for more effective living, at the other. For such an uninformed person, science begins with wondering or questioning, moves to the creation of possible explanations for the questions, and then ultimately to testing the validity of these explanations; technology is simply learning to manipulate nature for individual benefit, something man has been capable of doing since the earliest times; and sociology begins with the notion that no individual is an island, and that interaction with others is basic to human existence.

It is difficult to imagine interacting people who do not question, who do not wonder, who are not curious. It is hard to imagine a human who does not learn, almost instinctively, to use his/her surroundings for benefit (food, shelter, comfort, ease of movement, and communication). The interaction and interrelatedness of science, technology and society illustrate the problem inherent in school science as commonly conceived and practised. It is presented out of any real-world context. Its meaning and practicality are removed. It becomes a series of unconnected ideas and activities which teachers proclaim important, relevant and useful. But students are not permitted to see and to experience the ideas and activities in such a way.

STS means a complete focus on practical work, beginning with real-world questions as seen and identified by students from their own frame of reference. It means working on such problems. It means proposing possible solutions and explanations. It means collecting evidence concerning the validity of such solutions and explanations. It means presenting evidence about such validity that others will accept. Each stage is practical (meaningful and useful) and each stage is personal, which is to say that individual students are personally involved. All students have experiences with the natural world and invariably these experiences are more powerful than the ideas and the verification-type activities that characterise most science courses.

STS is the latest effort for producing students who are more literate in matters of science and technology following instruction (Miller *et al* 1980; Miller 1988; American Association for the Advancement of Science (AAAS) 1989). Miller's (1988) work continues to show that typical school science is ineffective in providing such literacy. One of his latest reports indicated that 94 per cent of US citizens are functionally illiterate in matters of science and technology. One explanation for this shocking failure of school science is the lack of meaningful practical work.

Recently NSTA (in press) updated its list of characteristics of a scientifically/technologically literate person. Such a person is one who, among other things, uses concepts of science and of technology and ethical values in solving everyday problems and making responsible everyday decisions in everyday life, including work and leisure; engages in science and technology for the excitement and the explanations they provide; displays curiosity about and appreciation of the natural and human-made world; applies scepticism, careful methods, logical reasoning, and creativity in investigating the observable universe; values scientific research and technological problem-solving; locates, collects, analyses, and evaluates sources of scientific and technological information and uses these sources in solving problems, making decisions, and taking actions; remains open to new evidence and the tentativeness of scientific/technological knowledge; and recognises the strengths and limitations of science and technology for advancing human welfare. STS efforts are designed to focus upon these personal characteristics as instructional goals. The curriculum is seen as a vehicle to develop such student characteristics and the strategies selected by the teacher to use the curriculum must also be directed to the production of students with such characteristics.

STS, with a broader view of science, makes it easier to identify specific dimensions which also provide a framework for evaluating the centrality of practical work as well as an indication of its effectiveness. Figure 3.1 is an attempt to provide a focus for setting goals related to the lives of students and to current problems as well as to a desired future. For many, work in the applications and connections domain may be all that can be done. To move to the concepts and processes of science before a student sees a need or a value will insure that learning meaningless information and recall for examinations will continue.

Many are now calling for the addition of applications to concept and process considerations. However, adding a third category without first hooking students on questions and their personal needs will probably not result in much real improvement. Most students will be lost – or turned off – before getting to applications. And in many instances teachers are uneasy with applications, which means that in practice there will not be adequate time to include them. Certainly applications of basic concepts and processes of most disciplines are not included in college courses. Perhaps new goals need to start with application – where information and processes affect student lives.

Nearly 1,000 seventh- and eighth-grade students (13- and 14-year-olds) were tested over a three-year period in certain Iowa schools where there were students in one class who had experienced science in a traditional manner and others in

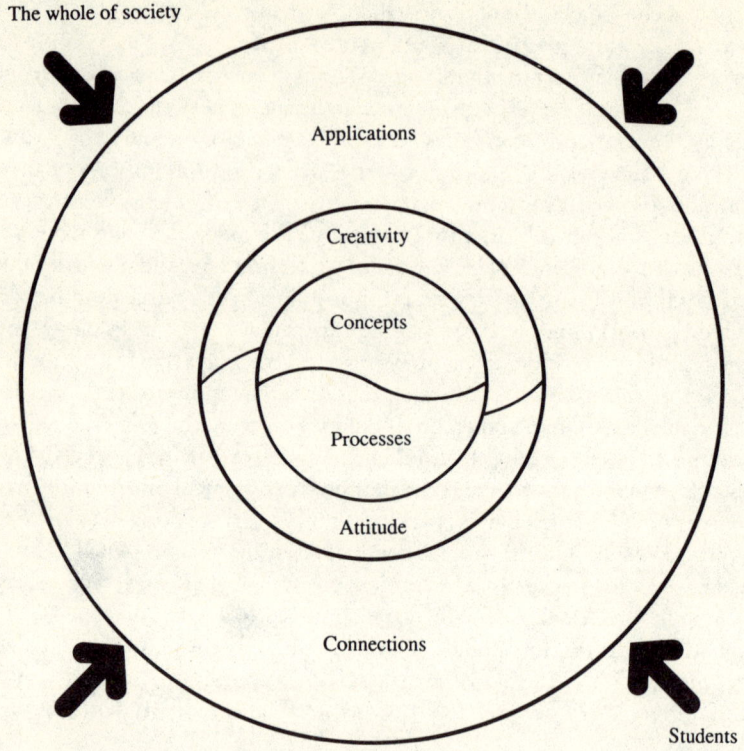

The whole of society

Applications

Creativity

Concepts

Processes

Attitude

Connections

Students

Figure 3.1 STS: domains for teaching and assessment

another class who experienced their science with an STS focus. The assessment efforts were all characterised by students being asked to perform in ways consistent with course objectives. Instruction was to be practical by definition. Five areas where students in STS courses excel illustrate the power of STS efforts for those valuing such student skill/learning outcomes. These areas include applications, creativity, attitude, process skills, and science concepts.

In the case of applications, each teacher was asked to identify an STS module that would coincide with a textbook or unit from the district course of study. For example, a toilet paper testing project provided the setting for considering the content of a month-long ecology unit. Twelve teachers were able to identify such parallel situations with at least one section randomly selected for the standard unit and one with an STS focus. The teachers devised their own assessment strategies for making specific observations of student ability to use information in new settings; to relate two happenings in a new situation; to identify related but divergent practices from a given situation; to choose relevant information for solving a

Table 3.2 Percentages of students who demonstrate ability to apply student learning in science, by teaching method

Percentage of students who can	Traditional teaching	STS teaching
Demonstrate the use of information in a new setting	25	81
Relate two phenomena to a new situation	18	66
Identify related but divergent questions from a given situation	17	83
Choose relevant information to use in solving a problem	26	91
Choose appropriate action based on new information provided	35	89

specific new problem; and to choose appropriate action based on new information provided. The same tests were given by each of the 12 teachers to each section of students who studied science directly as information presented and those who worked on problems with an STS approach.

The results are presented in Table 3.2. There were no observable differences in class sizes between the traditional and STS sections for the two sections of students used to assess in each of the five domains (class sizes varied from 18 to 32). There were no differences in abilities, male–female mixes, interest levels, grade averages, and other observable ways between the sections taught by the teachers.

One area where all teachers were asked to provide information was in the affective domain where items have been extracted from the released ones from the 1978 Third Assessment of Science by the National Assessment of Educational Progress (NAEP). Nine items from the Preferences and Understanding Instrument (McComas and Yager 1988) were identified as general areas of importance concerning student attitude.

Several studies have indicated that positive attitudes decline with the study of typical science. Generally student attitudes are more negative following the study of science for a given year than when students begin (Yager and Wick 1966; NAEP 1978a; Yager and Bonnstetter 1984; Yager and Penick 1986).

It was established before the instruction programmes began that the two sections of students were similar in terms of attitude in the nine areas and that these attitudes were similar to reports generally regarding student attitudes reported by NAEP. After two months of instruction students completed the attitude assessment for each teacher who had traditional and STS sections. Table 3.3 includes information indicating the different perceptions of students involved with each treatment who reported on the nine attitude indicators.

A third dimension of science learning and assessment focuses upon the creativity skills that are basic to science itself. These include curiosity about the natural universe and specific skills such as exploring, wondering and questioning.

Table 3.3 Percentages of students with specific attitudes to science, by teaching method

Specific student perception	Traditional teaching	STS teaching
Science classes are fun	40	81
Science classes are boring	31	14
Science classes make me curious	24	71
Science classes help me make decisions	31	63
Science teacher likes my questions	48	88
Science teacher admits to not knowing	22	74
Information from science classes is useful	69	81
Science is a favourite course	11	22
Science is a least favourite course	19	6

Creativity skills also include the formulation of possible responses and explanations for the questions raised and the planning and execution of tests for the validity of the explanations advanced. The creative skills define the basic ingredients of science and are the skills Torrance (1966) proposes for assessing the existence of creativity. STS teaching results in significantly greater growth in these skills and even greater uniqueness for some students than can be demonstrated in standard courses. Specifically, STS courses provide class averages that double such averages of classes taught traditionally in terms of number of questions, number of possible causes, and number of possible consequences. In terms of uniqueness of responses, STS classes are better than traditionally taught ones by a factor of six.

Process is a dimension of science which has received major attention in science education for over 50 years in the United States. Unfortunately, most of the attention has been lip-service, with little research evidence to demonstrate that science teaching resulted in students who possessed better science process skills than they had without instruction. Process items in each area of the processes of science identified by the AAAS for its *Science – A Process Approach* (1967) program for K-6 use have been developed over a five-year period. Teachers choose items for each skill most appropriate for their students and their content focus. Or they create new items most appropriate for use with their own students. Table 3.4 provides the results averaged from reports provided by the Iowa teachers.

Student acquisition of concepts has been the most common means of assessing success with the study of science in academic settings. Usually this focus has been one of exposing students to science concepts as presented in the textbook and/or by the teacher. Often the purpose of practical work has been providing direct experiences with the verification of some law or principle with actual materials. In the end success with instruction has meant seeing who can best repeat the concepts which were taught. Many have argued that the STS approach would result in fewer concepts mastered by students since the focus in the classroom changes from considering such concepts to dealing with questions and issues which require the concepts if responses and/or actions are to be taken. When studies are undertaken to compare concept mastery by students enrolled in STS classes with those enrolled

Table 3.4 Percentage of middle-school students who demonstrate their ability to perform in 14 science processes, by teaching method

	Traditional teaching	STS teaching
Selecting best experimental procedure	24	52
Hypothesising	18	63
Comparing and differentiating	31	84
Measuring	33	91
Using numbers	40	89
Predicting	19	71
Drawing conclusions	24	82
Controlling variables	21	63
Communicating	38	86
Inferring	19	74
Interpreting data	31	88
Classifying	26	87
Observing	30	84
Using space/time relationships	12	51

in standard classes, generally no statistically significant differences are found. In some instances some teachers have demonstrated that their students display more mastery in STS situations. However, the chief advantage of STS seems to exist in the retention of the concepts by the students. Learning concepts in a real-world context seems to result in longer retention of the concepts (Myers 1988; McComas 1989). And, of course, the 'quality' practical work found in STS settings may be responsible for the fine showing of STS students when comparisons of concept mastery are undertaken. It is noteworthy that direct coverage of concepts for the sake of coverage is not an efficient mode for achieving mastery.

Practical work is essential for successful STS teaching. However, such 'practical' work must be carried out from a student's standpoint as opposed to the standpoint of teachers and/or textbook authors. If such work is to be useful and meaningful, the student must identify and experience the idea, the activity, and the laboratory as needed. To teach basic concepts and processes and then to proceed with illustrating practicality and applicability is to rob students of direct experience. They are recipients instead of active participants. They forever prepare for doing science and technology; they never get to do it. Is it any wonder so many reject school science as meaningless, useless, and unconnected? School experiences in science which focus on the practical assume that most can see the importance of science, that most will improve in terms of literacy features, that most will improve with respect to their questioning, explaining and testing skills. In some respects practical work *is* science; it can be ideal science when it is a response to students' own questions, their explanations, and their tests of their explanations of events and objects from nature itself. This is quite a departure from an elaboration of science concepts currently accepted by scientists and practice with the process skills which they are purported to use as they develop new and more accurate concepts. So-called practical work which is teacher- or book-directed and which is primarily a matter of following directions and verifying what students have been told is not much of an improvement over learning concepts and process skills by rote. Quality science instruction means utilising quality practical work.

CHAPTER 4

Practical science in low-income countries

Terry Allsop

If science is to be learned effectively, it must be
experienced.   (Unesco 1973)

The laboratory is a unique facet of science education.
(Tamir 1989a)

Introduction

Although defining the countries to which this chapter refers is not straightforward,
they do have a number of factors in common: low per capita income; predominantly rural populations; economies based on primary products; relatively recent
experience of political independence; and limited access to schooling, particularly
at the secondary level. While they are situated on several continents, the majority of
examples in this chapter are taken from sub-Saharan Africa, the region which has
the greatest difficulties in providing the normal facilities of trained teachers,
laboratories and equipment for teaching practical science.

Context and aims

By the process of importing 'packaged' curricula, recipient countries unwittingly
submit themselves to a particular view of the world . . . There is an unspoken
assumption that the image of science described by the imported curriculum is
correct, and also that it is suitable for all pupils (Ingle and Turner 1981).

The world-wide period of centralised curriculum development which began in the
1960s has not left these countries untouched. Typically, as in sub-Saharan Africa, it
coincided with the granting of independence from a colonial power, but not with
freedom from the influence of the metropolitan countries. The models of science
curricula may vary considerably, but the continuing pattern of dependency – on
curriculum experts, on textbook publishers, on science equipment manufacturers,
and on examinations conceived thousands of miles away – remains hauntingly

similar. The examples are legion, even stretching to the influence of the science curricula of the Soviet Union on Chinese schools and universities in the 1950s. Perhaps the most notable feature of the curriculum development movements in science was the insistence on the importance of student experience of the science laboratory, usually through some form of teacher-directed investigational process. In the northern industrialised countries questions have only recently emerged as to whether this approach provides the most interesting or the most successful means of providing secondary school students with a liberal scientific education (Jenkins 1989). Examples of this form of cultural imperialism abound. For example, the Nuffield science projects of the 1960s, with their emphasis on student experimentation in the style of guided investigations, were taken up and modified for use in East Africa and Malaysia, in the former being grafted onto a tiny secondary school system still largely staffed by expatriates and resourced by imported equipment. In each case the later attempts to generalise this approach to an expanding secondary school population faltered, leading to headlines in the press such as 'Revert to Pure Science, Education Ministry urged' (*New Straits Times*, 5 June 1978). The way in which the Scottish Integrated Science course was adopted in only slightly modified form by a substanial number of countries, aided by the activity of commercial publishers, has been documented by Williams (1979). In taking on a tightly structured and well-tested course for junior secondary use, they were also buying a particular approach to practical science. Sim (1977) has reported that the enquiry approach espoused is at variance with the cultural values of many teachers and parents.

A pause for reflection on the aims of practical science which accompanied these curriculum projects may be appropriate here. In the context of northern industrialised countries, the most prevalent aims can be grouped under four headings: stimulating interest and enjoyment; learning experimental skills and techniques; teaching the processes of science; and supporting theoretical learning (Woolnough and Allsop, 1985). In espousing these aims, if only at second hand, countries took on commitments to the provision of resources at a level often beyond their realistic capacity to deliver, and, most important, the necessity of providing appropriate assessment instruments. Typical of such a commitment would be that made in the Zambia Education Reform Document (1977): 'Pupils should be able to master useful practical skills which they would apply in life in various ways. They should adopt a scientific attitude and approach, they should observe, collect information, draw conclusions and apply what they know.'

Present practice

What, then, may we expect to see of practical science in action? A trite answer would be 'With honourable exceptions, very little!', but usually with plenty of extenuating circumstances. It may be of help to consider in turn three generally recognisable stages: primary, junior secondary and senior secondary schools.

Science teaching in primary schools has, until recently, rather oddly, had a higher profile in the national curricula of many low-income countries than in those of the industrialised countries. This is presumably explicable in terms of the realistic expectation that many children will only receive primary schooling and therefore only have this opportunity to be acquainted with science learning. Many primary science curricula have been developed in Curriculum Development Centres which appear to have all the right emphases, for example, strongly community and environmentally orientated approaches, as in the splendidly conceived rural development projects of Namutamba (Uganda) and Bunumbu (Sierra Leone). The needs of teachers in initial and in-service training appear to be well served by Young's important book, *Teaching Primary Science* (1979). Practical science has been encouraged by the supply of kits of equipment to schools, and linked practical workbooks (for example, in Pakistan). Yet in thousands of rural primary schools, practical science simply does not happen. Why not? At least three crucial factors need comment.

First, very large classes, often with 60 or more pupils working in very cramped conditions, set organisational challenges in relation to the delivery of practical science which would deter most teachers. They parallel the difficulties of organising meaningful practical agricultural activities for large numbers of primary pupils.

Second, most primary teachers have a personally limited background in practical science and only slight confidence in teaching methods other than exposition. In many low-income countries primary teachers of necessity take additional employment to supplement meagre salaries. In consequence, they have little spare energy to plan for complex practical activities.

Third, those who complete primary schooling will often have to face a fiercely competitive examination for entry to secondary school, perhaps with only one child in 10 or 20 succeeding. This examination will rarely test science knowledge and understanding, and never practical skills in science.

At the other end of the school system, the senior secondary school, the typical picture is of a very highly selected group of pupils taking highly academic examinations in some cases still related to international examination networks, for example, the system of Advanced level examinations. Here, students' abilities in practical science may supposedly be tested through questions set in written papers (as in China), or through practical examinations set for all students (as is the practice of the West African Examinations Council). If the latter system is used, then a form of practical science *has* to take place in the schools, but is usually limited very closely to exercises which can be practised in preparation for the examination. Class sizes are by now smaller and so managerial questions do not loom so large, but resources now constitute a serious problem. The type of science curriculum prescribed normally assumes laboratory experimental work at a high academic level, dependent on the provision of equipment, chemicals and services which is often simply beyond the resources of the country. In a recent survey of 50 Zambian secondary schools, Varghese (1988) found that only three had adequate facilities for practical science, and even in these only demonstration experiments were performed by the teacher, equipment and chemicals being conserved for use in practical examinations. The three reasonably equipped schools were private foundations with alternative sources of funding not available to government schools. In the 1970s, splendid laboratories were built and equipped in many African countries with funding from the International Development Association (IDA), an affiliate of the World Bank. On a recent visit to Sierra Leone, the writer saw many of these in decay – one laboratory was being used as a poultry house, and in another all the chemicals and equipment were piled together in a heap in the preparation room while the laboratories were used as classrooms. A recent World Bank (1988) report concludes: 'Quality education is just not possible in laboratories and workshops that have no electricity or water because wiring, fuses and plumbing have deteriorated and where equipment does not operate because spare parts and consumables are lacking.'

In between, we find the political imperative to provide serious evidence of progress towards extension of access to secondary schooling at least for the junior secondary years, in response to the greatly increased numbers of pupils completing primary schooling. Again, science curriculum development has outstripped the provision of resources and trained teachers in many countries, particularly in respect of the predominantly rural schools which are traditionally under-resourced. One of the most innovative responses is the ZimSci course for Zimbabwe schools, where the quirk of delayed independence has allowed the developers to learn from others' mistakes and to develop a realistic, practically based course built around kits of equipment and associated text material made available to all schools. Even this project, designed for rural, day secondary schools, has been forced by public pressure simply to deliver in less expensive packages the same traditional science syllabus of established, elitist schools (Knamiller 1984). We shall return later to this dilemma.

Barriers to change

It has already been hinted that the prevailing paradigm of inquiry-orientated practical science may map uncomfortably onto the teaching and learning modes associated with instruction in the previously stable, pre-industrial societies of many low-income countries. Wilson (1983) has pointed out that an instructional mode where teacher and learner roles are clearly defined and differentiated according to expectations brought by the child from the home environment dominates most science lessons in African and Asian countries. I would wish to argue that this scenario particularly affects teachers of science, whose own experience of science simply reinforces and validates the transmission mode and expository style (not a problem unique to low-income countries!). Lee (1982), writing of an Islamic culture, concludes:

there also exist social attitudes and cultural traits in Malaysia which may be antithetical to the spirit of Western science. One cultural trait is that not only is it not right for the young to question their elders but teachers also share the same conservative attitude and are used to being directed from above. It is doubtful that this attitude of acceptance of authority can be conducive to the development of inquiry–discovery learning of science among pupils.

It seems to me that a more appropriate focus for our concern should be the ways in which science teachers are trained. At present the typical science teacher educator in a sub-Saharan African country has received his/her own science education in an elite secondary and higher education system, presenting an academic, rarified view of science which generates little of the openness of approach needed in guiding student teachers out of their own background. Frequently the same teacher educator has spent little or no time as a science teacher in primary or secondary schooling – the environment of college or university is demonstrably more attractive as a workplace. So the twin goals of developing an approach to practical science which is both open and enquiring but which also relates to realistic assessments of resources are very rarely seriously addressed, never mind attained. Somewhat bizarre mismatches may occur, as in Oman, where the linking of national educational aspirations with oil revenues has allowed the equipping of secondary schools with splended laboratories. Unfortunately, the teaching force remains entirely expatriate, recruited chiefly from Egypt and Sudan, from educational traditions where practical science is rarely part of the school experience. So far the laboratories gather dust, waiting for a well-trained first generation of Omani science teachers! Some encouragement can be gleaned from Koroma's (1975) study of Sierra Leonean science teachers, which showed that local teachers were more likely to have positive attitudes to investigational approaches than expatriate teachers, many at that time coming from the industrialised countries.

A final, and in my view very significant, barrier to be surmounted is the addiction to summative practical examinations as part of science assessments at the final stages of senior secondary schooling, typical of the Advanced level-equivalent examinations operating in most former British territories. They suffer from all the usual defects and limitations of nationally set practical examinations, exacerbated by the logistical difficulties for teachers in resourcing them, and, again for teachers, the very real pressures of maintaining security. While guaranteeing – the usual justification for their continuance – some degree of student practical science, they offer a very limited range of practical activity, with a very negative feedback effect on what occurs in earlier years in science. The continued existence of practical examinations precludes serious discussion and analysis of the purposes of practical science, but is staunchly defended by many indigenous science educators (the writer may be seen as an outsider here) who argue quite properly in terms of the rights of their secondary school students to have access to science education of an internationally recognised standard.

Appropriate responses

The search should be on for models of practice which can lead to national or regional versions of science education which relate, as King (1986) suggests is happening in India, to local traditions, development plans and modernisation strategies. Fully articulated models probably do not exist but some interesting experiments have come to light.

It is a slightly strange conjunction which brings to the attention of educators in both low-income and industrialised countries a concern for environmental, societally-based approaches to science education at this time. In low-income countries the argument is developed at least in part as a way of utilising the natural environment as a resource for practical science in the likely continuing absence of access to more formal laboratory science – we should probably view this positively. Not that this scrutiny of the environment is a particularly novel idea. Cole (1975), developing a course for Sierra Leone, was able to show that traditional African culture, properly employed, holds a rich source of materials for developing the 'scientific method' of enquiry and knowing about the various elements and processes in the African environment. More recently, Knamiller (1988) has, within a constructivist framework, systematically explored the potential of rural science and technology in Malawi to provide opportunities for students to extend their knowledge, to raise questions, to challenge current views and to learn skills of investigation. His conclusions are themselves very challenging to science educators: that it is relatively easy to devise investigations based on local science and technology; that it is much harder to generate among science educators enthusiasm for and *confidence* in such an approach; and that it is equally difficult to infuse community-based, investigatory science and technology into school syllabuses, teacher-training curricula and selection examinations at all levels.

If the debate is to be resolved in favour of this 'environmental' approach, I believe two questions have to be dealt with. First, is this approach intellectually coherent and sufficiently closely matched to the needs and aspirations of low-income countries, or will it come to be labelled as a second-class curriculum? Second, can the approach be properly applied in the rural primary or junior secondary school with very large classes and minimal resources?

The evidence available is very limited, and not all positive. The literature contains examples of exciting, one-off project activities, ranging from the study of traditional iron extraction methods in northern Uganda to a host of investigations of local alcohol production. There is the experience of Zambia, where a compulsory project became part of the national third-year secondary examination, since discontinued when the projects became stereotyped and almost entirely paper-and-pencil products. A similar project may be undertaken as part of the twelfth-grade chemistry examination in Papua New Guinea. A less radical approach is that advocated and worked out by Swift in his important book, *Physics for Rural Development* (1983). The starting point here is the given physics syllabus for Kenya, examined after four years of secondary schooling. The exciting aspect of the book, which is really a guide to teaching methodology for physics teachers, is that all the practical physics experiments are chosen with two criteria in mind – first, the context for the experiment is one which may reasonably be expected to be familiar to an average Kenyan student; and second, the equipment to be used should be available locally. (See the next section for a discussion of the Kenyan Science Equipment Production Unit.)

These examples are intrinsically interesting, but only provide indirect clues as to ways forward. Each of them has as a prerequisite a level of teacher confidence and competence which can not be assumed to be widely present. Although clarity of aims for practical science must be a national/regional priority, questions about necessary resources for delivery are very closely related, and will be addressed in the next section before a synthesis is essayed.

Providing the resources

Since the early 1970s, a substantial literature has grown up documenting attempts to solve the problem of major resource deficiencies for teaching science in nearly all low-income countries. It is piecemeal and often depressing because the contributions, although manifesting great enthusiasm, are frequently not integrated into a coherent policy for delivering practical science, and sometimes actually promote equipment development which is simply not appropriate. For *every* country, there is a range of ways of providing resources and equipment for practical science, which will include: improvisation by teachers in school; in-service workshops for equipment production; nationally produced equipment; and imported equipment. In a fully articulated system, all four elements will contribute to the provision of appropriate resources matching the needs of practical science in the curriculum. Each will now be considered in turn.

Improvisation by inventive teachers developing ideas for apparatus and experiments in their own work environment is hardly a new phenomenon – it is a very creative activity, traditionally close to the hearts of physics teachers in particular, and to famous scientists like Rutherford who coined the phrase 'string and sealing wax' to describe certain approaches to experimentation! The store of ideas in science teachers' journals round the world is tribute enough to teachers' ingenuity. But it is certain that it will not occur in those countries where teachers are unreliably paid and where they frequently have to supplement their income from outside teaching. The kind of creative use of improved equipment often seen in the primary classrooms of industrialised countries does not readily transfer to a primary teaching force lacking professional self-esteem, and to a society where one person's junk is the next person's artefact. Nevertheless, improvisation of simple equipment can be justified on a number of criteria, which have been well summarised by Simpson (1972):

1 It is cheaper, so that there is more apparatus available for individual or small group experiments, in addition to teacher demonstrations.

2 Concern over loss, breakage and repair is reduced, therefore equipment is more frequently used.
3 Students are made aware of scientific principles applied to everyday things, not just those associated with special apparatus imported from abroad.
4 Attention is drawn to the need to estimate accuracy.
5 Students can see where inaccuracies arise and can see the need for more sophisticated designs for many purposes.
6 A classroom can often be used if a laboratory is not available.
7 Simple equipment encourages students to make good use of local resources, and enhances self-reliance.
8 Simple experiments often demand an understanding of basic principles rather than the following of a set of complex experimental instructions.

Teachers working together in in-service workshops to produce materials to carry back to their own schools has, in the past, been a popular model for alleviating local equipment shortages. It has often related to the first flush of enthusiasm for practical science resulting from the introduction of a new curriculum. The benefits of such an approach are considerable – development of teachers' practical skills, production of useful apparatus, and camaraderie among teachers sharing professional expertise (and grumbles). However, the approach has limitations related to quality control, safety, and use of teachers' valuable time. Similar concerns surface when the in-service workshop is extended to in-school production of equipment involving students in the construction work. Undoubtedly the most sophisticated example of the genre has been that provided by Krishna Sane of Delhi University, who has targeted his workshop productions at a higher than usual age range, focusing his efforts on making chemical instrumentation for senior chemistry students. The designs and products include pH meters, colorimeters and conductometers. Sane's approach has been widely replicated, with evaluation suggesting that quality control can be maintained, and that product costs are of the order of 10 per cent of commercial equivalents for comparable performance (Sane, 1982).

Perhaps the most important merit of the workshop approach is that it has on occasion led to the development of local science equipment production centres. Many such initiatives have grown up since the early 1970s, on all the major continents. At their best, they reduce dependency on imported science equipment; they produce equipment of consistent quality; they involve teachers, teacher educators and curriculum developers at the design stage; they produce equipment closely related to contemporary science curricula in use in the country; and they are commercial enterprises. A common strategy has been to seek to develop kits of apparatus which can be given to schools as part of an integrated package with curriculum materials such as teachers' guides and student workbooks. The kits produced by Kenya's Science Equipment Production Unit (SEPU), which is based at the Kenya Science Teachers' College, contain equipment which allows the teacher to demonstrate crucial experiments. In Zimbabwe, the ZimSci kits provide for both teacher and student practical work. ZimSci has been particularly successful in utilising commercial waste materials, and in adapting for use as science apparatus such long-production-run items as measures, cans and plastic cups. These kits seem much more likely to be used than those sent to low-income countries from industrialised countries, which frequently gather dust. Even in successful production units like those mentioned, there remains the significant danger of copying from a manufacturer's catalogue while claiming to have made the product locally. And an approach through kits can be criticised as providing a form of packaged science which allows only the demonstration of simple and idealised phenomena, thus frustrating the exploration of real-life situations and problems (Ahmed 1977).

Choices

King (1986) has expressed the problem most clearly in the more general context of technological development for low-income countries, writing particularly of India:

> In the developing world, where development has seldom meant more than a mad race to catch up with the West, technological changes (in the North) will pose serious problems. If catching up with the West needed a major commitment in the 1960s and 1970s, it will require the total commitment of all our national resources in the 1980s and 1990s. This will raise serious questions of choice. Do we develop our science to stay in the technological race, to enter the 21st century on the terms of the world technological powers? Or do we develop our own science focusing on our land and water resources, on our forests and grazing and on removing the growing environmental imbalances that threaten the very survival of millions of our countryfolk?

Applied to our agenda for practical science education in low-income countries, this presents us with a number of difficulties. 'Science for all', perhaps broadly interpreted to include technology, is a legitimate aspiration, at least for primary or basic schooling in low-income countries. Acute problems arise when we try to interpret that for secondary schooling, where the familiar trappings of laboratories and equipment designed for small-group student practical science provide unrealistic expectations. Yet we have to note the necessity for low-income countries to develop the capacity to respond to relevant scientific and technological advances, such as the implications of genetic engineering in agriculture, through the skills of their own nationals. The World Bank (1988) puts it thus: 'Africa must improve its science and technology training and aim at the highest standards for at least a minimum core of specialists.' It is not clear whether the final part of the statement is directed at the level of secondary schooling – if it is, readers will have no difficulty in seeing the dilemmas posed for science curriculum planners.

For primary science, where there are already excellent models incorporated in programmes like the Science Education Programme for Africa (SEPA), a predominantly environmental approach (science studies in the environment, science at home, locally employed technology) is already possible with limited resources, but remains heavily dependent on investment in increased sophistication and confidence in the teaching force, and probably the recognition of such approaches in summative examinations at the end of primary schooling. Current investment in primary schooling in sub-Saharan Africa averages US$0.60 per pupil per year and it has been estimated (World Bank 1988) that a nearly tenfold increase to US$5 per year is needed to provide basically resourced primary schooling, with consequences for many countries of a rise in investment from 1–2 per cent of gross national product (GNP) to 3–4 per cent!

The case of secondary schooling is more complex. It can be argued that practical laboratory science is just part of a larger disfunctionality in secondary schooling, which relates to issues of credentialism. In such an environment it is unlikely that the aims of practical science discussed earlier will be fully operationalised. Practical science will continue to be seen as a luxury which no one can afford, except in the immediate run-up to a practical examination. Of course, we could make a case for examining practical science indirectly using paper-and-pencil tests which require the candidate to demonstrate planning, decision-making and problem-solving skills, but that has very serious logistical implications which have not really been solved anywhere. An interesting example of the beginnings of such an alternative approach has come from the work of regional groups of physics teachers in Cameroon. Their assessment was that it was quite unrealistic to expect students in the early years of secondary school to have an experience of practical science, given the usual prevailing shortages of resources and huge class sizes. They have prepared workbooks for students which aim to provide an indirect experience of practical science, sometimes through data analysis, sometimes through comprehension of descriptions of completed experiments, sometimes through 'thought' experiments'. The impact remains to be evaluated and no one is claiming that the approach provides a comprehensive experience of practical science. A radical approach would be to commend an extension of the environmental model from the primary sector to the secondary, matching King's implied preference, but inviting gibes of 'second class' and 'neo-colonial'! Otherwise the most optimistic view that can be offered involves the use of carefully designed, integrated packages involving kits of equipment closely linked with curriculum materials which allow the teacher to demonstrate experiments and to offer occasional practical experiences for the students.

Whatever directions are taken, the necessary sequence for implementation remains the same. First, clear national aims must be articulated for science curricula and associated practical science. Second, careful judgements need to be made about the reality or practicality of proposals derived for practical science. Third, it should be ensured that there is substantial investment in teacher preparation for delivery of proposals, relating teacher education curricula closely to school realities. Fourth, appropriate investment in facilities is necessary. Fifth, full recognition must be given to all genuinely creative local responses. And all this must be done by indigenous science educators!

PART III

The nature and purposes of practical science

A means to an end: the role of processes in science education

Robin Millar

Introduction: the aims of practical work

The characteristic of school science which most clearly differentiates it from other subjects in the curriculum is that science lessons take place in laboratories, where pupils and teachers carry out practical investigations and demonstrations. In the United Kingdom, 11–13-year-olds typically spend over half their science lesson time engaged in practical work (Beatty and Woolnough 1982), and 16–18-year-olds more than one-third (Thompson 1975). This emphasis on pupil practical activity, with its very substantial demand on time and resources, can be traced back in the UK to the Nuffield curriculum projects of the 1960s and beyond. But what is this practical work for, and what learning does it promote? Its very taken-for-grantedness means that this question is often not asked; we find it hard to imagine school science without a strong practical emphasis. We reply simply that 'science is a practical subject' and leave it at that.

When the question *is* explicitly addressed, answers are likely to identify two sorts of rationale for practical work: in facilitating the learning and understanding of science concepts; *and* in developing competence in the skills and procedures of scientific inquiry. Indeed, it has often been assumed that these aims can be achieved simultaneously, by the same sort of practical work. The Nuffield view of the 'pupil as scientist' (emphasising the processes of *doing science*) arose from a view of learning summed up by the maxim 'I do and I understand' (emphasising improved science concept learning). This underpinned many of the curriculum developments of the 1960s and 1970s in the UK.

More recently, however, these two sets of aims have increasingly come to be regarded as distinct aspects of science performance. Woolnough and Allsop (1985, p. 8) argue that a 'tight coupling of practical and theory can have a detrimental effect both on the quality of practical work done and on the theoretical understandings gained by the students'. They identify three distinct types of practical work: *experiences*, intended to give pupils a 'feel' for phenomena; *exercises*, designed to develop practical skills and techniques; and *investigations*, where pupils have an opportunity to tackle a more open-ended task as a 'problem-solving scientist' (pp. 43ff.). Where displaying and justifying scientific theory is the aim, they suggest that teacher demonstrations may be more effective than practical activities undertaken by pupils.

A recent Assessment of Performance Unit (APU) report on the performance of investigations by 13- and 15-year-old pupils also makes the distinction between pupils' understanding of science concepts and of how to carry out a scientific investigation. The APU authors conclude that 'performance on these tasks is influenced by both procedural and conceptual understanding' (Gott and Murphy 1987, p. 50), treating these as two distinct elements of understanding. This perception of *procedural understanding* – understanding how to 'do science' – as a distinct aspect of science performance has been consolidated by the establishing of *exploration of science* as a separate attainment target in the UK National Curriculum for science (Department of Education and Science (DES) 1989c), alongside *knowledge and understanding in science*, to be assessed and reported separately.

A process approach

Some recent reports and curriculum projects in the UK, however, have not merely identified procedural understanding as a *separate* aspect of science performance but have argued that it is the *most important* aspect – indeed, that it provides the fundamental rationale for science in the curriculum. DES (1985b, p. 7) asserts that: 'The essential characteristic of education in science is that it introduces pupils to the methods of science.' It continues by identifying the sort of practical work which this requires:

> courses provided should therefore give pupils, at all stages, appropriate opportunities to:
> – make observations;
> – select observations relevant to their investigations for further study;
> – seek and identify patterns and relate these to patterns perceived earlier;
> – suggest and evaluate explanations of the patterns;
> – design and carry out experiments, including appropriate forms of measurement, to test suggested explanations for the pattern of observations.

This view of the 'methods of science' as a series of discrete steps (or 'processes'), beginning with observation and leading, through classification to the drawing of inferences and the formulation and testing of hypotheses ('patterns'), underpins several recent curriculum development projects in the UK, notably *Warwick*

Process Science (Screen 1986a), *Science in Process* (Wray *et al.* 1987) and *TAPS* (Techniques for the Assessment of Practical Science) (Bryce *et al.* 1983). A particularly clear statement of the rationale for this approach is provided by Screen (1986b, p. 13). He argues that 'a knowledge-led curriculum has little relevance' but that there are

> qualities of science education which might be termed 'the primary or generic qualities' which will be of value when the facts are out of date or forgotten. If any qualities or generic skills are transferrable then the processes must form a substantial proportion, and any preparation of young people must take into account the transferable skills which they will need to succeed.

Warwick Process Science lists observing, classifying, inferring, predicting, controlling variables and hypothesising as the processes it seeks to develop, and organises its teaching materials into units with these titles. *Science in Process* uses the label 'process-skills' and identifies a similar (though not identical) list, which includes observing, classifying, predicting and hypothesising. Another recent course, *Nuffield 11–13 Science* (Lyth 1986, p. 17), lists observing, patterning and designing experiments as 'skills to be developed'. 'Patterning' is said to include classifying and predicting, while 'designing experiments' includes hypothesising. Despite these variations in these lists and in the associated terminology, there is clearly substantial common ground here.

Process-led, as distinct from knowledge- or concept-led, science courses are not, of course, a new idea. Although they have recently enjoyed a considerable vogue in the UK, they have been proposed and developed at other times in other countries, most frequently in association with moves to broaden the clientele for school science towards 'science for all'. Most can claim descent from the American Association for the Advancement of Science's (1967) *Science – A Process Approach*, based on Gagné's (1965b) view of 'science as process'.

The aim of such courses, as the quotation from Screen above makes clear, is to *teach* or to *develop* the so-called 'processes of science'. That is, the claim is made that very general skills can be taught, which can be transferred from the context in which they are learnt to new contexts and contents encountered later. The processes are not merely seen as the structuring elements of the curriculum, or as the means through which other more fundamental aims are to be attained, but as themselves forming the goals of school science.

In this chapter I want to argue that many of the explicit and implicit claims made on behalf of a process approach to science education simply cannot stand up to careful scrutiny (see also Millar and Driver 1987; Millar 1988). More specifically, I will argue, first, that it is misleading and unhelpful to portray the 'method of science' as a set of discrete processes; and second, that most of the so-called 'processes of science' are general cognitive skills which all humans routinely employ from birth, without formal instruction, so that it is absurd to claim that these can (or need) in any sense to be *taught* or *developed*. This second argument is the more significant. It argues not that we ought not to attempt to teach children the 'processes of science' but that it is impossible, in principle, to do so. For this reason I shall treat it at greater length.

Scientific method

Process science portrays the methods of science as a sequence or *hierarchy* of processes, beginning with observing and leading on, via classification, to inference and hypothesis. Thus observation precedes theory. This is essentially a *naive inductive* view of science. The central problem with induction is that we can never be sure that inductive generalisations are valid. The next instance we observe may, logically, refute the generalisation. Attempts by philosophers to produce a 'logic of induction' – a set of rules for making the step from a collection of discrete instances to a universal generalisation – have failed. (For a discussion of induction and its problems, see Chalmers 1982; Hodson 1982).

Popper's (1959) classic critique of induction centres around the asymmetry between proof and disproof. Although no finite number of confirming instances can ever prove that a generalisation is true (nor, strictly, even increase its *probability* of being true), a single disconfirming observation can, logically, falsify a generalisation. This led Popper to propose a *hypothetico-deductive* view of science. Popper argues that little can be said about the origin of hypotheses; the method of science lies in the rigorous testing of hypotheses in an attempt to falsify them. Only by falsifying an existing theory can we make progress towards a better one.

The hypothetico-deductive view has also become much qualified and elaborated in response to criticisms (Lakatos 1970). A central problem arises from the acceptance by most philosophers of science (including Popper) that *all* observations are theory-laden (Hanson 1958) – that is, what we observe depends on the conceptual apparatus, including the prior theories, which we bring to the task of observation. This means that a disconfirming observation is never 'pure', but depends on the theories of the observer. Perhaps it is *these* theories, and not the theory under test, which are wrong. So the idea that observation can conclusively falsify established theory is never quite so clear-cut (Mulkay 1979).

In the space available here, all that can sensibly be said about scientific method is that there is no consensus among historians, philosophers and sociologists of science. There are, however, some points of broad agreement. It is, for example, widely accepted that scientific enquiry resembles less the following of a set of rules (or hierarchy of processes) than the practice of a 'craft' – in deciding what to observe, in selecting which observations to pay attention to, in interpreting and drawing inferences, in drawing conclusions from experimental data (Polanyi 1958; Ravetz 1971). And the idea that observation is theory-laden is also widely acknowledged.

All this complexity is reduced, in the process approach, to a simple hierarchical model of science method. It might, of course, be argued that this really does not matter too much. The process view may represent the 'method of science education' rather than the 'method of science'. This is not, however, an argument used by the advocates of process, who typically appeal to the 'method of science' to justify their preferred approach. Let us in any event turn to the second of the two arguments outlined above, since it represents a much more serious and fundamental challenge to the process view of science education.

Teaching processes

The process approach implies that children require formal instruction in order to acquire, or to develop, facility in general 'processes' like observing and classifying. These very general 'processes', however, have no special link with science but are general cognitive skills and strategies which we all use all the time in making sense of the world. This may become clearer if we look at some of the proposed processes in more detail, focusing in particular on the question of what it would mean to try to improve a learner's performance on these processes.

Observing

We all observe. It seems strange to suggest that it is something we need to be taught. Of course, what is meant is that children should be taught in science lessons to observe closely, to notice detail, to make *relevant* observations. This, however, immediately identifies the problem. What is relevant? An observer needs some kind of prior expectation or theory in order to decide what features of any given situation are, or are not, relevant. By putting observation unproblematically first, the process approach disregards the insight that all observation is theory-laden.

Yet every science teacher 'knows' that observation is theory-dependent. Children's drawings of onion skin cells viewed down a microscope or of magnetic field patterns displayed by iron filings become very different once the child has been taught 'what to observe'. There is, however, no evidence that, once taught to observe in one of these domains, the child will be any 'better' at observing in the other. Instead, the evidence (Gott and Welford 1987) suggests that the quality of their observations in the new domain will be influenced by the concepts and theoretical ideas about the domain which they bring to bear on the task of observation and by the extent to which they are interested in the task, and not by their level of performance on some putative 'ability to observe' scale.

Some recent discussions of observation in science education begin from this perspective and try to explore the implications of accepting that observation is theory-laden (Norris 1983; 1984; Hodson 1986). Gott and Welford (1987, p. 226) propose that 'this theory driven view [of observation] should be adopted on educational as well as philosophical grounds. The consequence of this must be a re-evaluation of the role of observation in the learning and assessment of pupils.' The nature of this re-evaluation is more fully articulated by the same authors in a subsequent guide to the assessment of practical work in science (Gott *et al.* 1988, p. 37): 'We shall adopt the view of observation which argues that observation tasks are a way of ascertaining what conceptual understanding pupils have, or perhaps more accurately, what understanding they choose to bring to bear when faced with the task.' To illustrate this point, they discuss how children's recorded observations of a set of photographs of birds may vary, depending on the child's knowledge of the relationships between form (say, of beaks) and function. A child possessing such knowledge is more likely to comment on the beak shapes. Another may comment only on colour, or relative size. Thus the children's responses to the observation task give insights into the knowledge each brings to the task. The absence of a particular observation does not, of course, mean that the child lacks the relevant theoretical knowledge, so it is unclear whether observation tasks, seen from this perspective, could function satisfactorily in summative assessment situations, though they may be valuable for diagnostic purposes.

The objection is sometimes raised that surely some very basic observations are free of the influence of theory. But this, I think, is to misunderstand what is meant by the term 'theory' in describing observation as theory-laden. Certainly not all observation is influenced by theories that could reasonably be termed *scientific*: instead, many children's observations are influenced by 'common-sense theories' of one sort or another (Hills 1989, pp 175–6), or, in the cases of classroom tasks, by 'social theories' about what they believe the teacher *wants* them to notice and observe.

In all situations where we consciously observe, there is, as Hanson (1958) put it, 'more to seeing than meets the eyeball'. There is always much more information available to our senses than we can attend to. Even if we consider only the sense of sight, it is still the case that the visual field is so rich and detailed that we cannot simply 'observe everything'. We have to be selective, and this selection is inevitably guided by the theory or theories we hold about the thing we are observing and the observation task itself.

It makes little sense to speak of trying to improve someone's observation, unless reference is also made to the *kind* of observation we are speaking of – to the *purposes* of the observation. In science classes we may wish to train children to become better *scientific observers*. The way to achieve this is to extend the relevant theoretical framework they bring to any specific observation task and to present the task in a way which engages their active attention.

Classifying

Although the term 'theory-ladenness of observation' has passed into general usage, the idea of 'theory-ladenness of classification' is less usual. Very similar things, however, can be said of the so-called 'science process' of classification as of observation.

Along with observation, the ability to classify is the fundamental cognitive skill on which all knowledge acquisition depends.

> Human beings have the capacity to classify phenomena into groups. In principle every moment is novel, every situation is something that we have never encountered before, each object is slightly different. But in practice we do not see the world that way. We actually see the world as peopled by familiar objects, objects that we classify unproblematically and without question into conventional groups . . . The capacity to classify is something that we all possess and it is something that we do routinely . . . We have the capacity to notice similarity and difference (Law and Lodge 1984, p. 15).

The implicit (or explicit) claim in the process view of science that we should design certain activities 'to teach children to classify' or 'to improve their performance on the process of classifying' runs entirely counter to such a view. Yet, from birth if not indeed earlier, we all must be extremely adept at classifying if we are to make any sense of the natural and social world in which we find ourselves. Human infants quickly learn to recognise and respond to the shape of a face. As work on shape recognition by computer is making very clear, the cognitive task in classifying and identifying objects is an extremely complex one; so the infant's ability to make sense of the world by imposing the necessary order to develop stable object concepts is an astonishing achievement. To take another example, by the age of three, children can readily identify cats and dogs as distinct categories. Yet they would be quite incapable (as would most adults) of providing any usable definition of either category. Instead, their facility in classifying is based on learned perceptions of similarity, acquired through socialisation (see Kuhn 1977, pp. 309–18), but dependent on a 'given' propensity to classify. How can we sensibly claim that anyone needs to be taught to classify?

As with observation, the real focus for the science teacher must be on developing *scientific observation* – on helping children learn to appreciate the classifications which scientists use and their reasons for preferring to classify experience in this particular way. This is intimately tied up with the purposes of science. A whale is classified along with the mammals and not as a fish, not because this is obviously so, nor because it is necessarily entailed by 'the way the world is'. Rather it is a *decision* to give greater weight to the criterion 'bears and suckles live young' than to the criterion 'lives in the sea'. This decision, in turn, is based on wider theories which scientists hold about the origin and evolution of species. As Law and Lodge (1984, p. 19) observe: 'Unlike the *propensity* to classify . . . the *actual* classifications adopted appear to be *conventional*' (emphasis in original). It is important to recognise this conventional character of all classification schemes, including scientific ones. Scientific classifications are no better than many other common-sense classifications, and may indeed for some purposes be appreciably less useful; the justification for taking them seriously is inseparable from an understanding of the purposes and theoretical commitments of science.

Other 'processes'

Similar arguments apply also to the other so-called 'processes of science'. If learning is seen as an active process of constructing meaning, involving an interaction between existing mental schemas and new sensory inputs, and not as the mere reception of sensory data from 'outside', then it must involve not only observation and classification, but also hypothesising, predicting, testing hypotheses, and so on. It follows that the child must be able, from the earliest age, to perform all these processes at a high level of sophistication. These are available so early that it is difficult to avoid the conclusion that they are, in some sense, 'programmed in'. And the tasks in which the infant must successfully use them so as to make any sense of the world are so demanding that it is difficult to argue that they need, in any sense, to be 'improved' or 'developed'.

The challenge for science education is not to develop or to teach these processes. It is to present science in such a way that children feel that it is personally valuable and worthwhile to use the cognitive skills which they already possess to gain an understanding of the scientific concepts which can help them make sense of their world.

Beyond processes

Taking a critical stance towards the process approach in science education does not, it is important to stress, imply support for a return to traditional forms of 'transmission' teaching, in which knowledge is implicitly regarded as a commodity to be transferred from the head of the teacher to those of the pupils. Effective teaching in science requires that we develop activities which motivate and encourage children to make *use* of their skills of observing, classifying, hypothesising and predicting as a means of exploring and coming to an understanding of scientific ideas and concepts. The confusion at the heart of the process approach is between *means* and *ends*. The processes are not the ends or goals of science but the means of attaining those goals. We need to be quite clear, as Harlen and Jelly (1989, p. 40) comment in discussing approaches to primary school science that our reason 'for emphasising the use of science process skills is . . . that this is the only way in which [children] will build up useful ideas or concepts. We are not *teaching* the processes; we are *using* them to develop conceptual understanding. Additionally, I would argue, with Carey *et al.* (1988), that for science education to be effective we must also provide children with some insight into the purposes of scientific enquiry:

> process skills are important elements of a careful scientific methodology . . . Yet, the standard curriculum fails to address the motivation or justification for using these skills in constructing scientific knowledge. Students are not challenged to utilize these process skills in exploring, developing and evaluating their own ideas about natural phenomena. Rather, instruction in the skills and methods of

science is conceived of outside the context of genuine inquiry. Thus, there is no context for addressing the nature and purpose of scientific inquiry, or the nature of scientific knowledge (p. 2).

Conclusion: the aims of practical work reconsidered

I began this chapter by considering the aims of practical work in science. I have argued that the process view of science, in which practical work is regarded as a vehicle for teaching and developing high-level transferable cognitive skills, is untenable. Does this leave practical work simply as a means of teaching science concept knowledge, or can distinctive procedural elements of learning be identified?

A starting point must surely be the differentiation of the general category of 'practical skills'. Figure 5.1 shows one way in which this might be done. This chapter has focused on the *general cognitive processes* and has argued that these cannot (and need not) be taught. *Practical techniques*, on the other hand, *can* be taught. These are the specific pieces of know-how about the selection and use of instruments, including measuring instruments, and about how to carry out standard procedures. They can also be seen as progressive, in terms of both the increasing conceptual demand of certain techniques and increasing precision. The third category, *inquiry tactics* are perhaps best thought of as a 'toolkit' of strategies and approaches which can be considered in planning an investigation. These would include repeating measurements and taking an average; tabulating or graphing results in order to see trends and patterns more clearly; considering an investigation in terms of variables to be altered, measured, controlled; and so on.

In this model, progression in procedural understanding would be interpreted in terms of the techniques and tactics sub-categories. It would involve developing

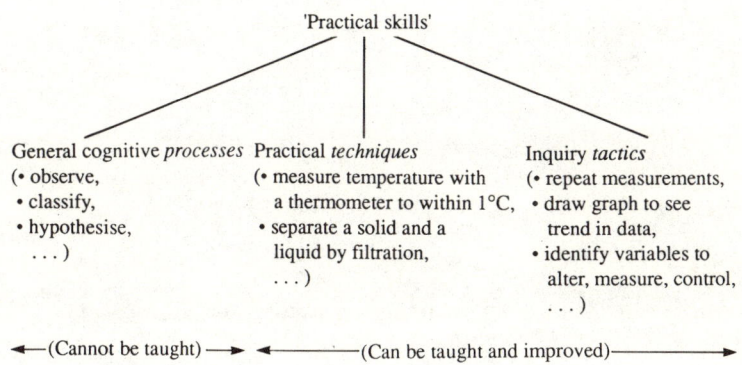

Figure 5.1 Sub-categories of 'practical skills'

increasing competence in a wider range of techniques, and enlarging and extending the 'toolkit' of tactics. The APU approach to investigation (Gott and Murphy 1987) focuses rather narrowly on the variables structure of an investigation which, in the model of Figure 5.1, is just one part of the tactics category. Much work remains to be done if we aspire to a model of procedural understanding comparable to those which we possess for conceptual understanding in many domains of science. One outcome of this work will be the discovery of whether such a model is possible.

CHAPTER 6

Practical work in science – a task-based approach?

Richard Gott and Judith Mashiter

Introduction

Science education is in a state of rapid change. In recent years there has been an ongoing debate concerning the importance of concept and process in our science courses. This debate has been brought into a tighter focus since the autumn of 1988 in England and Wales with the advent of the National Curriculum in science. At its centre is the role of practical work; while there is no great disagreement as to its importance, the argument concerns its aims – what pupils are to learn from their experiences in the laboratory.

In the classroom, the immediate and overriding preoccupation is with the implementation of the National Curriculum. Its emergence, together with its associated, but untested, national system of assessment, requires English and Welsh science teachers to reconsider the concept–process argument in the light of the restrictions it imposes and the opportunities it offers for innovation. It is being seen in some quarters as a heaven-sent opportunity to retain a concept-based course, while those who have set off down the 'process science' route are left in some doubt as to the extent to which they will need to cast aside an approach to which many are philosophically committed. To others it offers the framework for a genuine change. What is certainly true is that the urgency for change is now determined by the timetable for the implementation of the National Curriculum. The time is clearly ripe for a reconsideration of the science courses on offer, and in particular of the emphasis they place on practical work.

This chapter will not deal with specific examples of practical work. Rather it will argue for a curriculum based upon 'tasks' or investigative work which may or may not be practical (in the traditional sense of apparatus experimentation) but which involve the pupil in active data collection or analysis. The roles of practical work in various science schemes will be examined and will serve as an illustration of the argument being developed.

The social control of the curriculum

The authors believe that many of our problems in science teaching stem from our blinkered vision as to what constitutes 'science for all'. We wish to suggest that pupils' lack of success in science and their consequent disillusionment can be attributed, at least in part, to *our* failure to match the curriculum to their needs, preferring to deliver science courses that *we*, not they, feel comfortable with. We will begin by examining the underlying causes for that narrowness of purview.

The heart of most science courses has been the acquisition of scientific concepts. These knowledge-based concepts are often very difficult, abstracted from any real context, and – in the case of Nuffield schemes, for instance – embedded in an unreal world of ticker timers and trolleys that cannot be found outside the pages of school science equipment catalogues. As a consequence, pupils perceive no relevance to their science lessons other than the remote, and not necessarily very appealing, prospect of becoming the steriotypical white-coated scientist beloved of popular television advertisements. As Hodson (1987) notes, these 'high status, academic courses are characterised by their abstractness, emphasis on written presentation, use of individual rather than group work, competitiveness and unrelatedness to everyday life' and this curriculum 'represents a triumph of a particular interest group'.

The members of that interest group – the curriculum developers and opinion formers – operate from within a view of science which is based upon and attempts to emulate their own educational experiences which, at school and university alike, focused upon concept acquisition and explanation. Notions of investigative work which rely on the *use* of concepts rather than their acquisition or verification, are considered and discarded, or assigned to teachers of Technology or of craft, design and technology (CDT), or simply not considered at all.

This social control of the curriculum (perhaps subconscious social selection is a better description) does not operate solely on the 'content' – that is, the knowledge-based concepts – of science. The philosophy of science education and approaches to its implementation in the classroom are also largely governed by the historical inertia of inductivism. Most textbooks adopt this perspective of science as, arguably, does the Department of Education and Science (DES 1985b) policy statement, *Science 5–16*. The majority of Her Majesty's Inspectors' (HMI) documents, for instance, include a series of criteria which include observation, selection of observations for further study, and so on – an ordering which is driven, possibly unconsciously, from an inductivist position. This 'scientific method' derives from a retrospective examination of how scientists made great discoveries; it is by definition, then, tied to concept acquisition since this provides its evidence. It is, therefore, not surprising that practical work is seen as being concerned with illustration and verification.

Research evidence from the Assessment of Performance Unit (APU), as well as the everyday experience of teachers, suggests that this view of science has failed many of our pupils, particularly those of average and below average ability, in that 'a significant proportion of pupils appear to be unable to *apply* the scientific ideas that are being taught' in any other context than that used for their introduction (Gamble *et al.* 1985).

There is a need to challenge the prevailing culture which, because of its elitist view of science in schools, is condemning so many of our pupils to failure and frustration. We suggest that there are several key aspects to the problem of pupil failure in that the aims of the science curriculum are deficient in a number of ways. First, they are too abstract in terms both of the ideas themselves and of the contexts in which those ideas are usually taught. The difficulty of the ideas for many pupils means that their experience of secondary school science is one of repeated and demoralising lack of success resulting in a vicious circle of failure, demotivation and more failure. Second, they are not sufficiently motivating, in that they lack perceived relevance to pupils' own lives. Third, they rely on practical work as a means of enhancing 'conceptual learning rather than acting as a source for the learning of essential skills' (Fensham 1985). Finally, they are based (wrongly) on the premise that pupils will be able to use ideas, spontaneously, in a wide variety of situations. (The reader is referred to Fensham's (1985) reflective essay for a detailed and well-argued discussion of these issues.)

By examining the nature of the 'traditional' science curriculum and then considering some recent innovations attempting to change the curriculum and its teaching, an argument for a task-based approach will now be developed.

The knowledge transmission model

As we have already noted, the preoccupation in the 1960s and 1970s was with conceptual understanding, culminating in the influential Nuffield schemes. The pupil was seen as having a mind empty of preconceptions. The aim of science education was to fill that mind with the 'truths' of science. The scientist was portrayed, often quite explicitly, as a man (and very occasionally a woman) in a white coat in a 'high-tech lab' engaged in incomprehensible but, by implication, extremely difficult and important work. By studying science pupils could avail themselves of the opportunity to become members of this elite. Indeed, the course frequently suggested that pupils be allowed to become 'scientists for the day', as if being a scientist is something that one casts off before returning to the real world. The underlying philosophy of science was, and often remains, inductivist in origin. That is to say, pupils were introduced to a topic via experiments during which neutral observations were made. Hypotheses could then be generated from these neutral data and pupils would subsequently 'discover', or be helped to discover, the particular concept to be acquired.

With particular reference to the nature of *practical* work within this view, the purpose was to illustrate concepts so that pupils could 'see' them in action. 'The aim does not lie in the discovery process . . . [but in] the understanding of certain basic concepts' (Woolnough 1988a). In practice, the Nuffield approach became very didactic in the hands of teachers not well versed in its philosophy and faced with the broad range of ability in the emerging comprehensive schools. (For a fuller discussion of the Nuffield schemes see Woolnough 1988a.)

The pertinent point here is that all of this practical work had one aim only – that of introducing, illustrating or refining *concepts*. The consequence, in terms of pupil learning, has been that many pupils acquire fragments of knowledge which they can recall only in the context in which they were taught or reinforced. 'Good' pupils collect more knowledge fragments than their 'weak' peers. Opportunities to put science to use in relevant situations were limited since the practical work was so tightly defined and often used purpose-built and 'pupil-proof' (in theory!) apparatus.

The knowledge refinement model

Most courses aimed at knowledge transmission have an implicit view of learning centred on the premise that pupils have an empty mind into which a body of knowledge, called science, is to be transferred. An alternative view is based on the constructivist model of learning and teaching which emphasises the alternative frameworks, or naive preconceptions, which research suggests that pupils already hold (Millar and Driver 1987). Essentially, the alternative frameworks view supposes that pupils have their own perceptions of the environment involving language, beliefs and relationships and that these personal perceptions are frequently in conflict with the 'correct' concepts of the agreed body of scientific knowledge. Indeed, these perceptions are often referred to as preconceptions, as the intention is that they will be supplanted by the 'proper' concepts.

The constructivist approach focuses on the bringing about of change through revised teaching strategies but the overall objective is still the acquisition of the concepts that traditionally constitute school science. The challenge of teaching is then seen to be the effecting of change from preconceptions to the 'correct' conceptual understanding by providing experiences which either prevent pupils developing *mis*conceptions or which force them to confront the mismatch which exists between their ideas and the actual behaviour of the environment.

This approach suffers from two major drawbacks. First and foremost, its basic aim is identical to that of the concept approach – the acquisition of more, and more powerful, concepts. We are still primarily concerned with pupils' ability to explain, for example, the phenomenon of evaporation with the 'powerful tool' of the kinetic theory. Evidence suggests that many, and probably most, pupils are not able to do this. Whether alternative teaching strategies will remedy this remains a matter of doubt. Several years' experience of various programmes for accelerating development have produced little in the way of success. It could be dangerous to rest our hopes on this as a solution. And second, the approach still allows concepts to drive and steer the curriculum. All attention is directed towards concept illustration, the logical conclusion of which is the devising, often contriving, of methods, including experimental work, for doing this. The cleaner and more clinical the experiment can be, by the removal of the messy clutter which is reality, then the greater the chance of pupils' understanding the underlying idea. As a consequence, and with the best will in the world, the opportunity for choosing a context which is relevant to the pupils' experiences, rather than our own, is very limited.

Practical work can reveal the mismatch between pupils' preconceptions and the concept which is the desired learning objective. Thus its purpose is more than simple illustration and has more structure and direction than inquiry or discovery. Nevertheless, its role is primarily to facilitate the change in conceptual understanding and it therefore exists for and is driven by that collection of concepts to be acquired. This is not to suggest that such practical work has no value but rather to note its restricted and restricting role in the curriculum.

The point at issue here, as Fensham (1985) comments, is that 'the majority of the school population learns that it is unable to learn science as it has been defined for schools'. We think the problem is even deeper in that public perception of science almost *requires* that it be incapable of understanding; it only becomes science when it is not understood.

The 'process' model

Over the past few years several courses have emerged which have been given the label 'process science'. Often the 'process approach', as epitomised in such curriculum developments as *Science – A Process Approach (SAPA)* (American Association for the Advancement of Science 1967), *Science in Process* (Wray *et al.* 1987) and the *Warwick Process Science* project (Screen, 1986a), has been understood by teachers in a rather vague, ill-defined way as being somehow 'active', 'practical', skills-based and student centred. *SAPA* arose from work in the United States on a study of what eminent scientists do as part of their everyday activity. It is interesting to note, in passing, the reliance on this high-status view of science. Why do we not examine the use that 'ordinary' people have for science – the nurses, plumbers and intelligent laypeople?

Putting this to one side for now, let us examine the nature of a 'process approach' in more detail. The premise of the approach is that science should place more emphasis on its methods rather than focusing exclusively on its products. This is a challenge, at least in principle, to the accepted endpoint of science education as a body of conceptual knowledge. The processes include observing, classifying, describing, communicating, drawing conclusions, making operational definitions, formulating, hypotheses, controlling variables, interpreting data and experimenting.

In practice, the 'process approach' has come to have two strands. In the early years of secondary school science, the processes acquire a metacognitive role in which lessons are *about* 'observing', 'classifying', and so on. The processes assume the status of goals which mirror the concepts of the traditional or the constructivist approaches in importance. A lesson may focus on 'observation' and proceed to illustrate what it is to observe. But do pupils need to know that they are 'observing' or 'inferring'? The question as to whether or not metacognition will assist *pupils*, as distinct from teachers, to *use* the processes does not have an obvious answer. That does not mean that making such terms explicit to the teacher, so that they become part of teaching and learning, is not important. But perhaps they are better seen as a checklist to ensure that varieties of cognitive processes are included. Simply learning a definition, rather than being made aware of the 'correct' term for an activity with which they are already familiar, is unlikely to advance pupils any further in their understanding of science.

Typically, courses in the later secondary school years then deal with processes as a more efficient means to acquire concepts; they become the means rather than the ends of instruction. This is the viewpoint adopted by the constructivists who claim that processes are 'the vehicles by which children develop more effective conceptual tools. . . . For science, we would argue, is characterised by its concepts and purposes, not by its methods' (Millar and Driver 1987). So we find that the redefinition of science is only partial. While acknowledging that the methods of science are important, the methods are those of induction and operate within a concept acquisition framework. It is not likely, therefore, that the problems of relevance and motivation will be tackled effectively.

Procedural understanding in science – the missing element

Gott and Murphy (1987) have suggested that science is about the solving of problems in everyday and scientific situations. They chose to define a problem as a task for which the pupil cannot immediately see an answer or recall a routine method for finding it. To investigate and solve any problem, be it practical or not, there are a set of *procedures* which must be understood and used appropriately. In brief these procedures include:

- identifying the important variables
- deciding on their status – independent, dependent or control
- controlling variables
- deciding on the scale of quantities used
- choosing the range and number of measurements, their accuracy and reliability
- selecting appropriate tabulation and display.

Such activities are sometimes referred to as practical skills and as such are often taught in isolation. How many courses have 'circus'-type experiments involving the use of measuring instruments or the creation of tables, graphs and charts? Procedural understanding is not simply a collective term for such skills. The *Non-statutory Guidance* for the National Curriculum in science (DES 1989c) describes procedural understanding as an 'understanding of how to put all these specific skills together' via the identification and operationalisation of variables and the display and interpretation of data.

Parallels occur in non-practical problems. The idea of deciding which are the independent and dependent variables and which variables must be controlled applies equally to multi-column tables such as those found in *Which?*-type magazine reports (comparing the relative strengths and weaknesses of different makes of some commercial device) or in complex data on environmental effects, for example, as it does to planning and performing practical investigations; interpreting such tables often means 'cutting' the data – holding one or more variables constant, while looking at the effect of one variable on another. Data on the effects of pollution, for instance, can be accompanied by details of wind strengths and directions, maps showing the location of factories, farming patterns, and so on, which require a hypothesis to be formulated which can be tested, not necessarily conclusively, against the data.

We must beware of a temptation to think that such understanding is trivial. APU evidence suggested that, for instance, well over half of 13-year-olds were unable to manipulate two independent variables successfully (Gott and Murphy 1987). Pupils must have considerable practice before they can apply procedural understanding with confidence.

As we argued in an earlier section, most of the practical work in which we ask pupils to engage involves the illustration or 'discovery' of concepts. We ask pupils to follow instructions or 'do as I have done', fill in a table and display the data in a certain standard way. This is acceptable, even necessary, in experimental work designed to deliver a concept. But in a broader, task-based view, such a narrow conception of practical work must be rejected. The extra element of procedural understanding comes into play if this recipe of instructions is removed. It follows, as a natural corollary, that the science can be more 'open'.

The stratification of the science curriculum

To summarise the discussion so far, practical work has been used for a variety of purposes. Most commonly, it has been used to illustrate *concepts* on the basis that 'seeing is believing'. More recently, it has become fashionable to argue that practical work can be arranged in order to disturb and refine preconceptions – the idea of illustrating concepts has been replaced by the idea of active confrontation of preconceptions. The third approach, of 'process science', takes the elements of a 'scientific method' and elevates them to the same status as that given to concepts in the traditional approach. It may become necessary to add to that list the procedural approach to open-ended investigative work, which, through its influence on the National Curriculum, will begin to inform curriculum development in the short and medium term.

We wish to suggest that there is an underlying problem here of stratification which is distorting, even artifically polarising, the curriculum. In each of the above views, the key elements of science are extracted and taught in isolation. One stratum is represented by the powerful concepts of, for example, energy transfer. Another stratum consists of the processes of inferring, and so on. (Yet another could, in the future, be the procedures of investigational science).

Learning experiences, even models of learning itself, are then tailored to deliver these strata, often in deliberate and clinical isolation from each other. So, to take heat transfer as an example, experiments are selected *because* they will, we hope, deliver an understanding of the concept as efficiently as possible; the choice of context is secondary, at best. The process stratum, usually taught before the harder concept stratum, is designed to equip pupils with the tools necessary to deal with these concepts. On leaving school we expect the understanding from each stratum to merge, spontaneously, enabling pupils to solve a problem such as identifying and remedying those areas of a house which are not energy-efficient. The advantage claimed by this approach is one of economics; the ideas are applicable to any example in any situation, so we do not need to cover all situations, or indeed any of them. As Tobin (1984) points out in the context of a process approach:

> Process skills are not separated from science content when problems are encountered in real life situations. However, in classroom contexts, it is often convenient to isolate the processes so that activities can be planned to provide students with intensive practice on the skills.

The argument seems very logical and quite attractively neat and tidy. We know that it has worked for us, the controllers of the curriculum. But, of course, we have defined science from our inherited viewpoint. And, more importantly, the majority of pupils cannot share this perspective. Far from being powerful ideas for solving a variety of problems, science becomes an attachment of status and power for the few and fragments of disjointed knowledge for the many.

As Layton *et al.* (1986) note in connection with the historical development of science education and in the context of concept-driven curricula:

> Schools and colleges, in so far as they incorporated science in the curriculum, adopted a canonical version marked by abstraction and social disconnection . . . As science had succeeded in establishing a place in education for the people, so it had become insulated within the contexts of knowledge generation and validation and withdrawn from the contexts of use . . . In short, science for specific social purposes was replaced by science for science's sake.

Perhaps even more crucially, such a stratified approach denies the way pupils learn. It assumes that they will take on trust our assurance that all this knowledge will be useful to them, even as we deny them the opportunity to put it to use, there and then, in relevant situations. Such an abstract and deferred gratification is unlikely to motivate and encourage the majority of pupils. Indeed, evidence is accumulating which suggests that, to take one example, 'process skill learning appeared to be more effective when process skill lessons were infused into the regular science curriculum . . . rather than teaching the process in a brief topic as is often done' (Tobin *et al.* 1984).

A task-based approach to curriculum design

What sort of science should we be teaching? What role should practical work play? How can concepts, processes and procedures be brought together in a more relevant and effective manner? How is science to be made at once less abstract, more relevant and more motivating?

Our first step must be to re-examine the definition of procedures and processes, since there is some overlap. If we take procedures to be concerned with operations on variables (in a heuristic sense), then some of the 'processes' will come under such a definition, most obviously such ideas as 'controlling variables'. Those which remain under the process banner sit happily with the definition of Tobin *et al.* (1984) in which 'processes' are concerned with different modes of thought or 'intellectual operations involved when solving problems encountered in science and, more generally, in everyday life situations'.

The view to be advanced here is one which attempts to move away from a stratified and, to the pupils, fragmented, position to a holistic approach in which processes weld together and refine both procedural and conceptual understanding. We believe that the way forward is to construct a curriculum around a series of tasks that have within them the elements of motivation that stem from confidence in and a sense of ownership of the activity by the pupils. We would then have a model for science which engages all the elements we have discussed (see Figure 6.1).

It may be useful to consider an example of such an approach. The first requirement is to identify a context in which the task is to be set. For younger pupils, in primary schools (age 5–11), this may well be an everyday situation such as road safety. A task might involve them initially in investigating how Plasticine models are deformed in collisions with toy cars of various masses travelling at different speeds. It may develop these ideas using published data on accident statistics with a view to creating a news sheet about the dangers on our roads.

For pupils in secondary school a series of tasks may involve the investigation of structures for cars which compress on collision, or how one such structure crushes under different forces. (Other non-practical tasks may involve the collection and analysis of secondary data on stopping distances, manufacturers' data on the construction of collapsible box sections or steering columns, or data on the effectiveness of safety belts at different speeds and restraining bodies of different masses).

The role of 'processes', as defined earlier, now becomes apparent. The *processes*

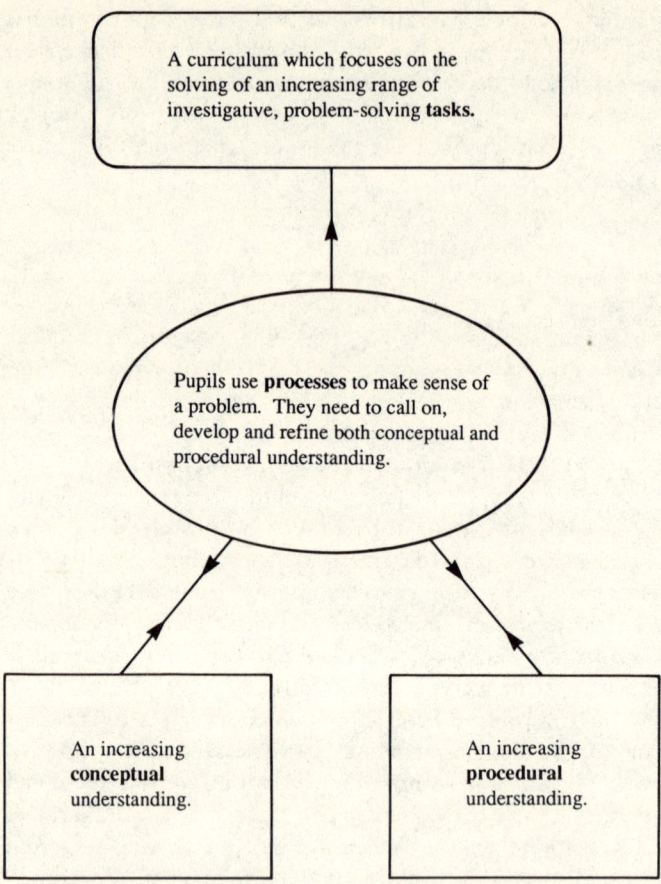

Figure 6.1 Processes mediate procedural and conceptual understanding in the solution of a task

are the various 'ways of thinking' that will be needed to co-ordinate the pupils' conceptual and procedural understanding into an overall plan for the task. As the task develops, they will *use and develop concepts* such as strength, force and deceleration *while utilising and refining the procedural elements* of the task – the strategies of deciding what to vary, measure and control and how to do it effectively to give valid and reliable results.

Would such activities be too hard, too abstract? Experience suggests that problems can be selected that are within the reach of the vast majority of pupils. It is true that there are differences in performance. Some pupils will carry out more sophisticated investigations in response to a particular task than others; but all will feel that they have achieved something. So motivation is enhanced because it depends so crucially on success. Moreover, the setting of the activities within meaningful contexts provides probably the best chance of encouraging transfer of procedures and concepts since the stratified elements of understanding are no longer divorced from purposeful problem-solving activities but are contained within them.

The key to such an approach would lie in the selection of appropriate tasks. They would need to be both practical and written. They would need to be selected for their intrinsic interest and motivation and would have to require pupils to draw on their increasing understanding of concepts and procedures, thus continuously refining them.

Progression within a task-based curriculum

Progression can be defined for a curriculum more easily than it can for assessment purposes. A curriculum, unless we are attempting to create an individualised learning package, can prescribe progression by reference to group behaviour, which is likely to be a little more predictable than that of any individual. Assessment, on the other hand, is by definition personal and subject to all the idiosyncrasies of the individual's interests, experience out of school, home background and so on, which we know to be so influential in pupil attainment. Given the limitations of space here, we will consider only progression in the curriculum.

The evidence to hand suggests that a large number of factors are influential in determining task difficulty. We consider three of the most important.

The first of these is *context*. Evidence exists from a variety of sources that the context in which a task is set may be one of the most important determinants of pupil success. Clearly a pupil with a particular interest in gardening, say, will be motivated and knowledgeable in a task involving the investigation of the effectiveness of weedkillers. Pupils' interests and background knowledge will broaden as they grow older, influenced by peer pressures and the immediate environment of home and school. It is against this increasingly complex and diverse contextual background that tasks should be set.

A second influential factor is *conceptual understanding*. Concepts interact with tasks in complex ways which require further investigation before we can begin to predict task difficulty with any confidence. Two examples may help to illustrate that complexity.

If the task requires pupils to work with concepts with which they have some acquaintance (heat and temperature, say), then they are likely to make progress – to the point where those implicit ideas can become clarified as they evaluate their solution against the demands of the task. If on the other hand, the concepts are not available even in an implicit form (electrical current, say), then pupils will have no basis on which to even begin the task, and what is more, little motivation to do so.

Once started on a task, the need to control variables becomes apparent in those circumstances where the pupils have sufficient understanding of the underlying concept to recognise its importance. For instance, in a comparison of the insulation properties of two thermos flasks, the need to control the initial temperature of the liquid relies on an understanding that the rate of cooling will be dependent on the instantaneous temperature. Many pupils who control other more obvious variables – the volume of the liquid, for example – fail to see the necessity for controlling the initial temperature. Add to this picture the fact that concepts themselves present gradients of difficulty which may well be pupil-specific, and the problems of progression through concepts become all too apparent.

A third factor is *procedural understanding*. Although gradients of difficulty in procedural understanding are largely a matter for conjecture in the face of a dearth of evidence, there is a small amount available from Gott and Murphy (1987) and Foulds and Gott (1988). We suggest that one way to begin the structuring of investigative work centres on the procedural aspects of the number and complexity of the variables involved.

Thus a *Which?*-style test involving independent and dependent variables both of which are categoric serves as a starting point (for instance, younger children, set the task of finding out which of two sorts of jelly dissolves most quickly may simply watch them and 'see which is fastest'). A second level would be a *Which?*-style test in which the dependent variable is continuous; we would be expecting pupils to quantify the rate of dissolving. Pupils would need to deploy the additional element of procedural understanding represented by the recognition and measurement of such a variable.

The next level would involve the move from categoric to a continuous independent variable. Pupils would now need to decide how many values of that variable to use, over what range and at what intervals: for instance, how does the rate of dissolving depend on the temperature of the water? The resulting data are best represented in a line graph, rather than the bar line chart appropriate to the preceding level.

Increasing the number of independent variables to two adds another level of complexity which, through research in the Piagetian tradition, we know to present difficulties to a significant minority of pupils at all ages in the secondary school phase (about a third of 16-year-olds). Pupils would be asked to find out whether the temperature *or* the size of the jelly pieces affects the rate of dissolving. The two independent variables would be categoric (the temperature of the water and the size of the jelly pieces, given two fixed temperatures and sizes, for instance) while the dependent variable, the rate of dissolving, would be continuous.

Once the idea of separation of these independent variables is mastered, the task can be changed by presenting pupils not with two sizes and temperatures, but with a packet of jelly and a source of heat. The task would ask pupils to determine how the temperature and the size of the jelly pieces affects the rate of dissolving and becomes a complex one involving them in decisions as to number and range of measurements for independent variables, both of which are now continuous.

Given this complex picture, how is task progression to be defined? We have argued in this chapter that a key issue in motivating pupils is that of relevance – choice of a context which sustains that motivation while catering for the development of the concepts that constitute a realistic curriculum is clearly, therefore, a vital element of curriculm design.

As to the more complex issue of procedural and conceptual understanding, we have suggested that to separate out the two would be a mistake; that to progress wholesale along first one stratum and then another is ineffective and demotivating. But that does not preclude an emphasis on one or the other in a particular task.

In structuring such a curriculum it may be useful to select an initial series of tasks, set in an appropriate context, in which pupils are asked to advance only along one of the strata (for example, working with a familiar concept, but advancing procedural skills). Once familiar with the procedural skills of *Which?*-style tasks, say, it may then be appropriate to deploy those procedural skills in contexts which involve less well-understood concepts – dissolving jelly might become linked to rates of reaction. The concepts can be developed within investigations, providing that pupils have some initial ideas which will allow them some purchase on the task. Other concepts which are very new to them may require an explicit introduction along more traditional lines before the ideas can be refined through use in investigative situations.

The trick will be to sequence the tasks so that the motivation derived from the relevance and completeness of the activity is not subverted by the need to control progression in concept and procedure, whether that need derives from assessment purposes or the structuring of schemes of work. Such a curriculum is, at present, not much more than a gleam on the horizon, but still a gleam worth striving towards.

The National Curriculum – an opportunity and a challenge

The National Curriculum in science introduced in England and Wales, revolves around two Profile Components (PC1 and PC2). The first of these has within it just one Attainment Target (AT1, exploration of science) while PC2 has 16, all of which are concerned with elements of knowledge and understanding. As with the other subjects, the science proposals have met with a mixed and surprisingly muted response. To some they represent a backward step into yet more concept acquisition. Certainly the sheer number of attainment targets in the knowledge and understanding Profile Component is somewhat forbidding. But there is more than a glimmer of hope. In Key Stages 1 and 2 (ages 5–11) something close to half of the curriculum in science is to be devoted to AT1, the fraction reducing to about a third in Key Stages 3 and 4 (ages 11–16).

AT1 has been construed by many as being simply practical work. This is not the case. Much of what we have called illustrative practical work in fact finds its place in the knowledge and understanding attainment targets; which is where it rightly belongs, since its purpose is to help demonstrate concepts. AT1, by contrast, is concerned with procedural understanding as we have defined it here. In the words of the *Non-statutory Guidance* for science (DES 1989c), 'this understanding of the way in which skills are subsumed into the tactics and strategy of an investigation is referred to as procedural understanding and is fundamental to Attainment Target 1'.

There is, however, an inherent danger in the separation of the two profile components. By definition, these components are separated for reporting (that is, assessment) purposes. But they may well influence the curriculum in such a way that it becomes stratified, with all the consequences we have identified in this chapter. It will be very tempting for schools to modularise their science curriculum and 'do' AT1 in a concentrated lump before moving on to the 'real' science. The *Non-statutory Guidance* for science advises strongly against this move. It recommends that AT1 should permeate the entire science curriculum, to say nothing of design and technology where a similar philosophy has been adopted (see Kimble, this volume). This permeation is, in our view, best achieved through a task-based curriculum. From this perspective, the progression defined in the Statements of Attainment, with the Profile Components seen as existing in isolation, represents an incomplete and rather simplistic view of pupil attainment but one which, none the less, gives a starting point.

To date, little work has been done to discover either how effective such a task-based curriculum would be, or how pupil attainment can effectively be described. Where tasks have been used in small-scale studies, the evidence, such as it is, suggests that the motivation and success of the pupils has been improved (Foulds and Gott 1988). More research must be carried out into the creation of banks of tasks, structured and indexed upon their procedural and conceptual complexity. Only then will we be in a position to develop and test the ideas presented here.

It is our hope that if this challenge is met, the science curriculum may become available and enjoyable to all pupils, and not just to the most able.

Acknowledgement

The authors would like to thank Patricia Murphy of the Open University for extensive comment and discussion during the writing of the early drafts of this chapter.

CHAPTER 7

Reconstructing theory from practical experience

Richard F. Gunstone

'Constructivism' has become a significant word in the lexicon of science education. Its importance derives from the plethora of recent studies of students' conceptions of science-related ideas and phenomena. These studies point to the personally constructed nature of these student conceptions. There is now considerable evidence to support assertions such as: learners enter science classrooms already holding personally constructed ideas and beliefs relevant to what they are required to learn; these ideas and beliefs are often at odds with the tenets of science, and remarkably resilient in the face of classroom teaching; in some science areas (such as mechanics) the ideas and beliefs show considerable consistency across age and culture of learners, while in other areas (such as health and sickness) there are major cultural differences in the ideas held by different groups; the conviction with which some of these personal constructions are held and used makes the descriptor 'personal theories' eminently reasonable.

At first glance it seems most tempting to see the changing of contrary student conceptions, the restructuring of ideas and beliefs so that these become ideas and beliefs of science, as being achieved simply by using practical work to show directly the inconsistency between the contrary conceptions and science. This view suggests that using the laboratory to have students restructure their personal theories is straightforward. However, such a view is a naive one. The real picture is more difficult, and the messages for practical work to be derived from constructivism rather more complex. These messages are the focus of this chapter.

The constructivist position on learning is taken as given. Readers unfamiliar with this position, and the research underpinning it, may like to consult one of the considerable number of reviews and books now available (a variety of perspectives are represented by Driver *et al*. 1985; chapters by Driver, Gunstone, Shapiro and White in Fensham 1988; Osborne and Freyberg, 1985; West and Pines, 1985; White, 1988).

This chapter begins with a real example of a teaching sequence which included practical work and demonstration, and which was aimed at having students reconstruct contrary personal theories. That is, both the planning for and intended outcomes of the sequence were rooted in constructivism. The purpose in giving this example is to use it to point to issues I see as significant for practical work.

A teaching sequence which includes practical work aimed at student reconstruction

It is important for later arguments briefly to describe the context of the teaching sequence, and all of the section of the teaching sequence in which the practical work is embedded. It is also important that previous teaching and research had led to an understanding of the likely personal theories to be found in this group of students (science graduates in a one-year pre-service teacher education course) for the content involved (direct-current (DC) electricity). These expected personal theories were broadly similar to those found among high-school students (see, for example, Osborne and Freyberg 1985). The sequence draws on previous work with school students (Shipstone and Gunstone 1985).

The example is the first five hours of a teaching sequence given in 1989 to 32 Diploma in Education (directly equivalent to PGCE) science graduates at Monash University, in the Australian state of Victoria. The graduates were all chemistry or biology majors, with many having passed first-year university physics. In their teaching careers in Victoria, these students will be required to take general science between the seventh and tenth grades as well as their specialist subjects to senior levels. Thus the content of the teaching sequence, fundamental concepts in DC electricity, includes concepts these students will teach in schools.

The sequence had two broad purposes: to develop an understanding of DC concepts and, subsequently, to use the examples of pedagogy exhibited in the sequence as a way of considering alternative teaching and learning approaches for secondary school sciences. The Monash science teacher education course of which this sequence was a very small part has a strong constructivist orientation throughout both its curriculum and teaching. The sequence occurred some months into the course. Hence the use in the sequence of pedagogies such as interpretative discussion (Barnes 1976) was not novel to the students.

Pairs of students were given a 1.5 volt DC cell, some pieces of wire and an unmounted torch (flashlight) globe. The task was to get the globe to light. A very small minority found the task trivial. A much larger minority found great difficulty and completed the task only via mimicking successful groups.

The purpose of this small task was to provide concrete orientation for the written work which followed. Each student completed a series of written questions about a drawing (*not* a diagram using symbols) of the circuit in the exercise just completed. The questions, adapted from Osborne and Freyberg (1985, p. 25), required yes/no answers together with reasons and probed ideas about presence of electric current in the globe, in each wire, and magnitude and direction of any perceived current in one wire relative to the other. These questions identified the models (personal theories) held by the students about electric current in a simple circuit. Of the four models identified by Osborne among school students (Osborne and Freyberg, 1985, pp. 23–4), only two – constant current and diminishing current – were present. It is very rare to find either of the other two models – clashing currents and only single wire needed – among science graduates.

A Predict–Observe–Explain (POE) task relevant to the written questions was then given. In this approach to the use of demonstrations, students are shown a particular situation and asked to predict with reasons what they believe will happen. Written predictions and reasons are particularly valuable because of the commitment this requires from students. The demonstration is then performed, and each student writes what he/she observes. Finally, reconciliation of any conflict between prediction and observation is undertaken. (A more detailed description of POEs and their purposes and uses is given in Gunstone *et al.* 1988.) Here the POE involved circuit the students had just used, with the addition of a demonstration ammeter in series in each connecting wire. The ammeters were described only as meters for measuring the size of the current in the wire. The prediction asked for was whether the reading on 'meter 1' (the meter in the wire from the positive terminal of the cell) would be greater than, equal to, or less than the reading on 'meter 2'. About 60 per cent predicted equal readings with all the rest predicting a higher reading on meter 1 than on meter 2. When the demonstration was performed some saw equal readings and some saw a slightly higher reading on meter 1 than on meter 2. There was a high correlation between prediction and observation. Then an intriguing event occurred. You will recall that this was a large group (32), so that only those near the front had a good view of the precise readings on the meters. Some students spontaneously got up and moved for a closer inspection – all of these having predicted a higher reading on meter 1 than on meter 2. Other students near the rear of the room reclined back in their seats with an air of contented boredom – all these had predicted equal readings. A consensus observation was accepted by the group after discussion: meter 1 did read higher than meter 2, with a difference of about 3 per cent of the meter 2 reading. Some, but only some, of those predicting equal currents saw the very small difference as a function of the meters. Others predicting equal readings had a rather more superficial understanding of relevant concepts, and hence became confused. Confusion was also present among some who predicted a higher reading on meter 1 than on meter 2, particularly among those who expected a more dramatic difference. I did not make the obvious statements about school meters (which these were), nor did I make assertions about what 'should have happened' (those students who predicted unequal readings had had such assertions made to them many times before in their extensive science learning). Resolution of these issues emerged during the stages which followed. We did discuss at some length relationships between personal ideas and observations.

An anology was then used to try to foster student discrimination of energy and current in the circuit under discussion. (The analogy was that of a bicycle: the chain as 'energy carrier' remains 'constant'.) There was also discussion of the specificness of the analogy, that is, that this analogy should be used only for consideration of this issue. The dangers of extending analogies beyond the specific purpose they serve were also discussed when energy and current were separately considered later. The analogy and discrimination of concepts helped many (but not all) resolve the constancy of current issue.

Then came a focus on energy issues. Briefly this involved debate, teacher assertions, discussion of examples to establish various points: a gravitational potential energy analogy: the circuit discussed above can be seen as an energy transformer: a volt is a joule per coulomb (why this should be useful remained unresolved until the chemistry of DC cells was later addressed); a mechanical model for resistance and heat production (resistance described in terms of 'collisions'); insulators and conductors in terms of energy transformation required to have detectable charge flow. Notes summarising these issues were handed out, and homework set which aimed to foster student processing of the notes and begin student thinking about the chemistry of DC cells. This particular session finished with the requirement to write down 'two questions which are puzzling you'.

Extensive discussion of the puzzling questions (a few of which still related to the issue of the constancy of current) and of the 'chemical logic' of using a measure of joule per coulomb rather than joule then took place.

After this, the concept of current became the focus of attention, again via discussion and consideration of questions and examples. A water-flow analogy was introduced to elaborate nature and flow of current, effect of resistance on flow, measuring current, conventional current. The logic of a relationship between potential difference (voltage) and current was established, and Ohm's law given. Both in notes then handed out and verbally, it was explicitly stated that Ohm's law did not describe every context. This particular session concluded with a concept-map task for homework. (Concept maps are described and discussed in Gunstone *et al.* 1988; Novak and Gowin 1984.) Terms to be included on the concept map were current, electric charge, electrical energy, potential difference, heat, circuit and resistance.

A practical exercise was then done. The practical involved varying the voltage across a torch globe and measuring current. Students were instructed to record data for voltage and current in tabular form, graph these data and draw a conclusion. Students worked in groups which I deliberately made large (three or four students). All the resulting data looked curved rather than linear to me. Nevertheless, linear graphs were very common (over three-quarters of all students produced linear graphs). A small number of these linear graphs were 'lines of best fit' drawn in surprising fashion over points I saw as clearly showing a smooth curve. However most linear graphs came from one of two other approaches: non-linear scales on the horizontal axis, or deciding to not plot any of the points for voltages at which the globe filament was not white hot (thus giving a very approximate linear graph). One group of four, all beginning with a single set of observations, had each of its members generating a different graph.

Most conclusions were of the form 'we have verified Ohm's law as far as it is possible with this equipment'. The practical gave a dramatic illustration of the importance of personal theories. Here it was not only the observations which were directly influenced, (as was the case in the written questions described above), but also the acceptance and interpretation of observations. Again, the issue of personal Theory–Observation–Interpretation links was discussed at length with the group. As I expected, this discussion generated considerable feeling. Many students felt that they had been tricked (that it was reasonable to expect a practical experiment of this sort, in this context to 'verify' Ohm's law). A practical which aimed to extend in the way of this example was unreasonable. Late in the discussion there were also confessions about the way that some had interpreted their undergraduate practical work to mean that knowing the expected answer before beginning, and then ensuring data which gave this answer, was an appropriate strategy.

The rest of the teaching sequence is not described here.

Before turning to a consideration of constructivist messages for the purposes and nature and use of practical work, two general comments about the example are important. First, the students involved in the teaching sequence example were unusual targets for this content on two counts: they were all apparently successful products of many years of science learning; and they were all highly motivated to try to come to an understanding of the concepts, a motivation deriving from their knowledge that next year they may well have to teach the same concepts. Second, the example shows clearly and unambiguously the impact of students' personal theories on their approaches to and learning from practical work (an impact which, as argued below, involves more than personal theories about science-related phenomena). This is a major factor in the need to see practical work in the context of a teaching sequence, an issue also well illustrated by the example.

Constructivist messages for practical work

There are many issues with which one might begin this discussion. I begin with observation because of its utterly central role in any teaching approach which aims to have students use real events from which to learn.

Observation

A series of related aspects of observation are relevant. For convenience, these are considered separately.

(1)   Observations are theory-dependent. Students' personal theories affect what they see. This is starkly illustrated by the POE in the example above, and also shown by the practical exercise. This in no way denigrates observation. Certainly I have no sympathy with the view sometimes argued that science is somehow absolutely relative. But the reality is that '"Looking at" is not a passive recording of

an image like a photograph being produced by a camera, but it is an active process in which the observer is checking his perceptions against his expectations' (Driver 1983, pp. 11–12). More colloquially (and, perhaps, more cynically) 'seeing is believing – I wouldn't have seen it if I hadn't believed it' (Brilliant 1979, p. 40). It is clear, then, that one cannot assume that all in a class have seen the same thing (even if all have looked at the same set of apparatus), that what the teacher sees is what students see, and so on. The beginnings of the implications of this, and points (2) and (3), are discussed in point (4) below.

(2)   Students' personal theories can lead them to reject observations via a denial of the legitimacy of the observations. This is illustrated by those students in the practical work in the example above who, in their desire to verify Ohm's law, ignored all data taken when the filament of the globe was not white hot. There are many other examples to support this assertion. Sometimes these contain some complexities. Take, for example, a POE in which a piece of rotting liver is placed in a jar which is sealed by double dipping in wax and then weighed. Students are then asked to predict with reasons the relative weight of the jar in two weeks' time (commonly predicted to be less, because rotting equals disappearing or gases have no weight). When the jar is found to have the same weight, some students will conclude that the jar was not airtight after all – same weight means air has got in (Gunstone *et al.* 1988). Other examples show more clear-cut rejection of observation. A POE used with 470 first-year physics students by Gunstone and White (1981) involved a block of wood and bucket of sand placed at rest at the same level on either side of a bicycle wheel mounted as a pulley and able to turn freely. The block of wood was pulled down (and held), and students asked to predict with reasons what the block would do when released. About half predicted that it would move up to its initial position. The observation of no movement on release was made by all. However, most of the students faced with the problem of reconciling this observation with their prediction of return to original position did so by denying the observation (commonly by arguing that the block was held for too long at its new position, so it 'got used to a new equilibrium position'). One cannot assume, then, that, even if all in a group see the same thing, all in the group will accept the validity of the observation.

(3)   The inferences drawn from observations are influenced by personal theories. Again this is clearly shown by the practical work example above, particularly by the group of four who produced four separate and different individual graphs from a single set of data. So, even if all in a group see the same thing and accept that this is a valid observation, one cannot assume that all will interpret the observation in the manner you have in mind.

(4)   Despite the crucially central role of observation in practical work, students rarely understand issues such as the nature of observation, what claims may validly be made as a result of an observation, what is and how to undertake the distinction between observation and inference, what purposes are intended in undertaking particular observations, and so on. The most obvious reason for this lack of understanding is, of course, that it is rare for these issues to be explicitly addressed

in science teaching. The teaching sequence example given above raises some of these issues in an indirect way. Here is a more obvious example, again taken from Gunstone and White (1981). Another POE task given to the same group of first-year university physics students required prediction, reasons, observation for a falling blackboard duster dropped from about 2 metres. The focus was on the relative speeds at a marker 1 metre above the ground and just before hitting. In some ways the most disheartening result was that all students gave an observation of relative speed at the two points. I must confess that, when I observe this event without any observational aids such as time-interval measurers, I cannot see a difference in speed at the two points. That all these students did shows something of their lack of understanding of the nature of observation (which I have already suggested is not their fault), and the development through their science learning of an expectation of making a 'correct' observation (again, hardly their fault). This interpretation is reinforced by the number who gave quantitative observations such as the speed at the lower point is about one and a half times the speed at the higher point. One extreme individual claimed to observe that the speed at the lower point was 1.414 (presumably $\sqrt{2}$) times the speed at the higher.

The essential message to be derived from this discussion is the obvious one: we need to help our students learn about and understand the nature of observation if we wish to successfully use a pedagogy so centrally dependent on observation. Driver's (1983) discussion of constructivism and observation is an excellent source. Readers who wish to explore this area further would do well to read her account. One helpful approach to the specific issues described here in points (1) to (3) above is shown in the teaching sequence example. That is to use chosen examples of observation and subsequent discussion to help students realise the effect of their own theories on their observing and inferring from observing, the importance of discriminating between observation and inference, and the claims which can validly be made from observation. The POE strategy is a powerful approach here because the use of predictions with reasons can so readily bring out personal theories prior to observing.

Student understanding of the purposes underlying the making of particular observations is equally important. This is totally intertwined with student understanding of the purposes of particular practical work, a fundamental issue which is discussed in a later section.

(5) There is a fifth aspect of observation to which constructivist perspectives point, an aspect that is not addressed by the teaching sequence example. It is, unfortunately, relatively common for observation to be used in classrooms to 'verify' principles which are not observable. Consider, for example, attempts to 'verify' Newton's second law. In order to undertake such a task, it is clear that one needs to be able to measure resultant force; to do this one must be able to measure friction; *ergo* it is not possible. Of course, there is great learning value in using laboratory trolleys to investigate relationships between mass, acceleration and *applied* force in low-friction and different friction contexts. The learning difficulties which can arise from unreasonable attempts to 'verify' observationally are well

illustrated by another example, one which I confess is a particular hobby-horse of mine. This is the generalisation that all objects have uniform acceleration in the earth's gravitational field. This is an idealised and conditional statement which can be 'seen' to be correct only for a small range of objects over a very small range of distances, and then only because of the inadequacies of the eye as an instrument and the hands as release mechanisms. Philosophically it is inappropriate to claim to verify this generalisation observationally. Constructivist psychology points to a more pragmatic objection. It is true that heavier objects fall faster than lighter objects in an atmosphere. This has been tried and seen by many of the students we teach. If we claim physically to demonstrate equal accelerations, and thus make claims quite counter to previous observations made by many of our students, we are being naive to think that our claim will cause students to reconstruct a 'heavier-falls-faster' personal theory. We will only do this by helping students integrate air-resistance effects with Newtonian perspectives and their prior experiences, and even then we will not be successful with all students. A related issue in this particular case is the common but erroneous claim that Gallileo observationally 'verified' this generalisation by dropping things from the Leaning Tower of Pisa.

In discussing points (1) to (4) I have given only a small number of examples. There are many others to be found in constructivist-orientated investigations of student learning (see, for example, Gunstone and Champagne, 1990, for different illustrations of some of these issues). There are also examples to be found elsewhere in descriptions of science itself (see, for example, an anthropological study of a leading immunological research centre by Charlesworth *et al.* 1989, in which some of these issues recur frequently).

Qualitative laboratory exercises

Traditional practical work has features which can inhibit the possibility of students restructuring personal theories. In particular, the tasks of assembling apparatus and making required measurements can themselves become the prime, or even sole, focus for student actions. In these circumstances, student reflection on the nature and implications of the observations is extremely rare. The practical work example above again provides an illustration. For many students the connecting of the required circuit was a major and very difficult task (a point not previously mentioned). For these students, successful assembly of the apparatus became the only significant task. Once this was achieved the rest of the practical was completed in a ritualised fashion, with little or no serious thought.

The essence of this issue is that, for practical work to have any serious effect on student theory reconstruction and linking of concepts in different ways, the students need to spend more time interacting with ideas and less time interacting with apparatus. By sometimes using shorter and more conceptually focused practical tasks, usually qualitative so as to remove the measurement task obsessions which consume so many students, the teacher has a much greater chance of fostering student *thinking* about the conceptual relationship(s) of concern.

POEs are one very useful possibility here. Although it might be argued that a demonstration is not practical work, I see an appropriate POE as generating much more student activity (intellectual) than does the extreme case of the recipe-following use of apparatus. Apart from POEs there are many possibilities which are more genuinely practical work. I give one, which is relatively complex, and which focuses on the qualitative relationships between applied force, mass and acceleration. Have students take two laboratory trolleys, one loaded with bricks. Connect a rubber band to each and place the other end of the bands through a metal rod (e.g. the vertical arm of a retort stand). Now, with the metal rod perpendicular to the direction in which the trolleys will move, pull on the rod. The motions of the two trolleys will differ, as will the applied force on each (shown by the extension of the rubber bands). Eventually each trolley will have the same acceleration. Considerable discussion is needed to establish, first, a comprehensive and common set of observations, and second, a Newtonian interpretation.

Purposes and examples of qualitative laboratory exercises have been discussed elsewhere (Clement 1982; Champagne *et al.* 1982; 1985). There is considerable similarity between this idea and Woolnough's notion of 'experiences' (see, for example, Woolnough and Allsop 1985, pp. 56–9).

Student conceptions of teaching, learning and purpose

It is not only science-related student conceptions (personal theories) which impact on learning from practical work. Even greater influence comes from student conceptions of teaching and learning and purposes of classroom activities, and from student beliefs about their roles in these processes. At the beginning of Chapter 8 of this book, White describes a student who sees the purpose of a practical as no more than the completion of a sequence of numbered steps. The example, from Tasker (1981), is only one of many in that chapter.

Consider again the practical exercise in the example teaching sequence (above). On completion, but before the discussion of graphs/conclusions and the influence of personal theories, students wrote responses to 'how does the experiment link with what we did last session?' A very small number of responses were insightful. Not surprisingly, given the account in 3(h) above, many saw the link as 'verifying Ohm's law'. Some could do no more than state that 'electricity' was the link! That is, most students either saw a purpose for the practical that was quite different from that of the teacher or saw no specific purpose at all. This is a major factor in so many of the students not learning what the teacher intended.

This common lack of understanding of purpose often derives from student conceptions of teaching and learning. As Tasker (1981) and others (for example, Baird and Mitchell 1986) have shown, students often do not even accept that they should understand purpose. Conceptions of what teaching and learning ought to be also impact in a major way. Consider, for example, discussion, a strategy I have already mentioned often as a necessary component for conceptual change from practical work. If you teach students who think that learning involves writing notes and that a teacher who cannot give an understanding to them is inadequate, then discussion will have no value until their conceptions of teaching and learning change so as to value this process.

The importance of these student conceptions cannot be overemphasised. The second half of White's contribution to this book (Chapter 8) elaborates aspects of these student conceptions and their impacts on learning from practical work. Hence the issue is not discussed further here.

Metacognition

Two overarching themes run through the previous discussions. First, it is fundamental to learning for students to be personally and actively linking the concepts, experiences and broader contexts which they possess and which are involved in the practical work they undertake. Second, it is crucial for students to understand and control the how, when, what and why of such linking. Metacognition refers to the understanding and control of one's own learning. Since all students have some form of understanding of learning and some conceptions of their roles and responsibilities in terms of control, even if these are of the inhibiting and unfortunate type alluded to above, then all students are metacognitive. The challenge is to develop in learners enhanced metacognition, an acceptance of their need both to understand and to take control of their own learning.

Any serious attempt to elaborate approaches to the enhancing of metacognition would require the major part of another chapter, and, again, the second half of White's chapter in this book explores aspects of such approaches. Hence I merely point to examples of other sources here: for further discussion of the nature and importance of metacognition in school classrooms see Gunstone and Baird (1988); for strategies for promoting enhanced metacognition see Baird and Mitchell (1986), Mitchell and Baird (in press).

Conclusion

As noted in the previous section, dual themes run through much of this chapter: the importance to learning of linking, and the need for students to control their linking. It is these themes which justify the need both to see practical work in the broader instructional context and to have students understand better the processes of practical work, particularly if the focus is on conceptual restructuring. Not only must teachers see practicals in the broader context, so also must students and in a manner which reflects enhanced metacognition. Hence the use of strategies such as POEs, concept maps and interpretative discussions is argued here. The use of problems and apparatus which are part of the everyday world of the student is also an obvious approach to encourage linking.

Student discussion, so often referred to above, must be purposeful in terms of promoting considered reflection by students of the content and purposes of learning tasks. And, for students to gain what we intend from the approaches argued here, they must be both willing and able to undertake such reflection. We have learned much in recent years about how to help students be able; sadly, we are a long way from understanding what might make a student willing.

CHAPTER 8

Episodes, and the purpose and conduct of practical work

Richard T. White

> Tasker: What have you decided it is about?
> Pupil: I dunno, I never really thought about it . . . just doing it – doing what it says . . . it's 8.5 . . . just got to do different numbers and the next one we have to do is this (points in text to 8.6). (Tasker 1981, p. 34)

This brief scene from a secondary school science laboratory indicates that practical work may be no more than the mindless following of directions, and that little useful learning is likely to occur during it. If that were to become widely believed, practical work would vanish from school science. After all, laboratories cost a lot of money. They require special furniture and extensive plumbing and electrical services, and are expensive to equip and maintain. Also, practical work is costly in teacher effort, since the time that teachers spend in preparing and supervising laboratory sessions could be given to direct teaching of many more students. So although teachers, curriculum constructors, educational administrators, and the parents and governments that provide the money still appear to have faith that practical work adds a dimension to learning beyond what can be gained from listening to a teacher or even from observing demonstrations, it is timely to see whether a sound justification for practical work can be supplied.

Unfortunately, there is no consensus about the purpose of practical work, or what might be the special dimension of learning that it fosters. One attractive notion is that experiences in the laboratory are necessary, and possibly sufficient, to bring students to exchange naive beliefs about phenomena for sophisticated scientific views. So, for example, the laboratory should convince students that Newtonian principles of dynamics are more accurate than Aristotelean ones. Unfortunately, it has been found that beliefs persist even in the face of direct contrary experience (Gauld 1986; Gunstone *et al.* 1981).

Another notion is that the laboratory provides training in problem-solving. There are different opinions about this: Gould (1978) found that teachers had come to accept it, while Osborne (1976) obtained contrasting views from undergraduates, who saw the laboratory rather as a medium for creating interest in science. That view, in turn, was contradicted by tenth-grade students surveyed by Ben-Zvi *et al.* (1976b), who rated promotion of interest last of a set of eight aims.

Differences in views may suggest that there is little point in further attempts to specify the purposes of laboratory work. However, recent years have seen substantial progress in the development of learning theories, which enables more precise discussion of the role of the laboratory and what might be done to make it a more effective learning experience.

One of the most crucial advances in learning theory is discrimination between types of knowledge. This is illustrated by a comparison of Ausubel's (1968) and Gagné's (1965a) writing about learning. Where Ausubel concentrated on meaningful verbal learning, that is, the acquisition of coherent bodies of propositions, Gagné wrote about the achievement of intellectual skills, tightly defined algorithmic procedures for carrying out classes of tasks such as balancing chemical equations or identifying particular species. The distinction between types of knowledge is important because each type has its own conditions for learning, and each may contribute in a particular way to the comprehension of scientific concepts and phenomena.

The questions set in almost any examination, in science and other subjects, require performances that draw on propositions and intellectual skills. Courses of study and textbooks tell the same story: propositions and intellectual skills are dominant in our formal systems of education. They are not, however, the only sorts of knowledge. Gagné and White (1978) and White (1988) draw attention to images, episodes, strings, motor skills and cognitive strategies as further forms that are important determinants of the quality of learning. Of these forms, episodes are especially important in considering how practical work contributes to that quality.

Episodes are recollections of events. All of us have many episodes in our memories; indeed, we tend to use the word *memory* to mean our collection of episodes. It is strange that so central a part of mental functioning has been ignored by formal education, or at best been dimly appreciated through the survival of excursions, laboratory work, and other forms of active learning. The importance of episodes in learning is that they are powerful elements in the understanding one has of any concept or situation.

Understanding of a concept or phenomenon is a function of the extent and mix of knowledge the person has about it. Consider, for example, the concept of chemical change. A student might have acquired knowledge of propositions, such as 'mass is conserved in chemical reactions' and 'physical changes are more easily reversed than chemical changes'; intellectual skills, such as being able to complete equations for classes of reactions; images of the rearrangement of molecules; and episodes of carrying out changes, such as burning of magnesium or pouring acid on a carbonate. The student's understanding of chemical change depends on how well integrated this knowledge is – whether it is a collection of unconnected elements, or whether it is all linked into a coherent whole.

The quality of a person's understanding depends also on the proportions of the different sorts of knowledge. One person may know many propositions about chemical change but have few algorithms to hand, while another has the reverse. Both have a degree of understanding of chemical change, but the qualities differ. The proportion of related episodes that the person has also determines the quality of understanding. This is well described by Gordon (1976, p. 223):

> Once one has watched the Brownian movement one's apprehension of the *nature* of heat will never be the same again. It is not that one can be said to have learnt anything in an objective scientific way but rather that one has come to terms with the kinetic theory of heat at a subjective level. It is the difference between having a sunset described and seeing one.

Gunstone and I (White and Gunstone, 1980) became aware of how understandings differ when we interviewed science graduates on their knowledge of electric current. Some respondents gave us fact after fact, but had little to say when we asked them whether they had any personal experience with electricity. Others told us tale after tale of experiences with rewiring motorbikes, touching spark plugs on cars, replacing house fuses, and so on, while reporting fewer facts. It is debatable whether one of these styles of understanding is better than the other, but they certainly differ. Probably most people interested in the learning of science accept that a balance of propositions, intellectual skills, motor skills, images and episodes represents better understanding than an extreme proportion of any one sort of knowledge. Few, at any rate, could object to the principle that episodes add to understanding.

The justification of laboratory work is that it can be a prolific source of episodes that learners link to the propositions and intellectual skills that are acquired in more teacher-centred lessons. It is not enough, however, for students just to do something in the laboratory for an episode to be formed. People store as episodes only a fraction of the many events they experience each day, and even if an episode is formed it need not be automatically a useful one, illuminating other knowledge and making it more comprehensible. The design of laboratory events must therefore take into account how humans process experiences.

Information-processing and constructivist theories of learning provide insights into the formation of episodes. Accounts of such theories applied specifically to science are given by Osborne and Wittrock (1985) and White (1988).

The first step in the mental processing of an event has to be its selection for attention. We simply do not notice much of what happens around us. Unless we 'put our minds to it', we ignore distant traffic sounds, the texture of pavements under our feet, or the colour of distant buildings. Selection of attention may not seem to be a problem with laboratory work, but an observation by Atkinson (1980) is disconcerting: a ninth-grade class had been taught about the gas laws, and the main feature of the lesson was a demonstration using a large cylinder and plunger; when the apparatus was shown again to the pupils *the next day*, about a third said they had never seen it before. That, of course, was a demonstration, and perhaps it would be more difficult to remain as unconscious in active laboratory work. We should, however, think of designing experiments so that they focus attention.

Selection for attention is only the first step in the processing of an event. Next comes making sense of it, possibly forming emotions about it, perhaps classifying it, and even deciding to remember it. People do this for few of the events they experience. Consider the events of the day – getting, dressing, having breakfast, travelling to work, and so on. How much of this do we bother to remember? Most of it simply flows through our consciousness without being stored in detail in memory. At best what remains is a generalised script, an outline programme of what one does each day.

How can laboratory experiences be arranged so that they are processed by students? In a study of my own memory for events (White, 1982; 1989), I found the no doubt unsurprising result that I remember vivid and rare events. (There were surprises in this research, for instance that it did not matter how central a part I played in the action nor how important the event was to me at the time.) We should, then, try to devise unusual and striking incidents in the laboratory. Certainly these do occur, but in my experience they have mostly been unplanned accidents from which little useful learning resulted – an explosion or two, a fire, dropping and breaking a bottle of oil. It would be good if practical work could be planned to form useful episodes as clear as those I have for these accidents.

It is easy enough to see what makes an event rare, but what makes it vivid? Presumably there has to be something remarkable about it. I once saw carbon monoxide ignited in a large gas jar; the event remains vivid for me because of its surprising beauty: the flame was a delicate blue and took a paraboloid shape, and as it descended in the jar a musical note was emitted that changed smoothly in pitch. There was magic in this for me. It might be difficult to devise vivid experiments, but we should try. Many now common in schools are humdrum, and merit little more attention than Tasker's pupil gave to step 8.5.

An example of what might be a vivid physics experiment is one I devised for transformation of energy (Gardner and White, 1972). Students make a catapult out of rubber bands, and find out how much work is done in stretching it by constructing its force–extension graph and finding the area under the line. Then they use the catapult to fire 50 gram masses at $45°$ to the horizontal; the measured range allows calculation of the kinetic energy at release. The fun of firing the catapult should make this a memorable experiment.

The need for vivid, rare experiments does not mean that everyday ones are useless. If they are processed, their repetitive sameness will produce generalised episodes or scripts that still are useful forms of learning. Thus, although you might not be able to recall a specific instance of measuring something, repeated experience of reading meters, graduated cylinders and rules leaves you with knowledge of the procedure of measuring, including the skills of taking care and the awareness of parallax that are involved in attaining accuracy. Repeated experiences teach you how to behave in the laboratory. They give you familiarity with chemicals and apparatus, and with abstract concepts such as energy. They can also teach you the fundamental principle of scientific method, of holding all variables but one constant in order to see the effects of that one. Repeated visits to the laboratory transmit an impression of science, and have a strong influence on attitude to science.

Simply visiting the laboratory and following instructions without thought, like the students interviewed by Tasker, will not result in useful learning of either specific or generalised episodes. The experience will pass and no imprint will be left. It must be processed. The problem in conducting practical work is how to engage the students to that they do reflect on what they are doing.

Episodes will readily be formed for vivid and rare experiments. However, almost by definition, only a handful of experiments in a year can be rare, and vivid ones are hard to invent. Therefore, teachers need procedures for more everyday (or every-week) practical work that will encourage the students to think about what they are doing. One recommendation is for the laboratory work to be concerned with real problems, for which no procedure is told initially to the students. They have to think about the problem, and write out the procedure for the teacher to check. They will take time to learn how to do this, so the early problems they get should be simple ones. For example, the problem might be 'Does a hole in a piece of metal get bigger or smaller when the metal is heater?' There are many ways of seeking an answer to this, which is good since students should learn that science has alternative procedures. A criticism of the way science is currently presented in schools is that there appears to be only one way of doing things, or of conceiving of things.

When students have learned how to investigate simple problems devised by the teacher, they can move to more difficult ones and some that they propose themselves. In Bentley and Watts (1989), Heaney, Doherty, Naylor and Watts describe four successful implementations. The success appears to stem from the relevance of the problems to the students' interests (for example, reasons for cleanliness in a McDonald's fast food store) and their ownership of the problem. Relevance and ownership guarantee processing.

A second recommended way of engaging students in thinking about procedures is applicable to even the standard, cookbook-type exercises that are so common today. I owe it to a biology teacher in the Project to Enhance Effective Learning at Laverton High School, Melbourne. Concerned about her students' mindless following of instructions, she merely provided the steps for the next laboratory exercise in a jumbled order. Before the students could start the exercise they had to think about the steps and get them into a sensible sequence. This simple trick threw responsibility for actions on to the students. The teacher found that they accepted that responsibility, thought about what they were doing, and incidentally reduced the times that they came to her with questions about managerial matters that they could work out for themselves (for example, where to find test-tubes).

Episodes must not only be formed from practical work, but to be truly useful they must also be linked by the learner to the propositions and intellectual skills that make up the bulk of the subject matter of the topic. Linking is not certain to occur spontaneously, although many practices assume that it is. For instance, it used to be common (and in some places may yet be) for a class to have a regular period of practical work in which there was one piece of apparatus for each of a ring of experiments. At any time there might have been as many different experiments going on as there were students. No experiment was done by all the students until the whole cycle was complete, therefore the teacher was inhibited in referring to any experiment. Consequently there could be little attempt by the teacher to get students to associate the experiment with other knowledge. Under the worst forms of this arrangement, which I experienced myself as an undergraduate, changes in the lecture syllabus are not co-ordinated with changes in the laboratory pro-gramme, so that experiments continue on topics that are no longer taught, and for which the students have no knowledge to link their new episodes.

Even when all students do the same experiment at the same time, and that time is appropriately close to the other teaching of the topic, they may not link the episodes that they form with relevant propositions and intellectual skills. When Moreira (1980) asked undergraduates what key concepts were relevant to an experiment they had done, few could say. Tasker (1981) also found that secondary school students saw little connection between practical classes and other science lessons.

It should not be too difficult to devise ways of encouraging students to link laboratory episodes with propositional knowledge. For instance, the students could be required to list in their report on the experiment all the propositions they used or thought of during it. For the well-known experiment (described in Haber-Schaim *et al.* 1971, p. 34) in which two carts are driven apart by springs, so that momentum changes can be compared, the report might list:

Explosions are the sudden release of stored energy.
I did work compressing the springs to store the energy.
Moving masses have kinetic energy.
Some kinetic energy is lost through friction.
Momentum is a vector.
Momentum is conserved in all directions in an interaction.
If a big piece goes in one direction and a small one opposite, the small one will go faster.
Hand grenades are grooved so that bits will fly off in all directions.
(Why aren't shells and bombs grooved? To lower air resistance − so are they grooved inside?)
The centre of mass of the carts remains stationary.
The centre of mass of a shell that explodes in the air keeps going, so more pieces fly forward than back − that's why infantry can move up behind an artillery barrage.
An astronaut who drifts away from the spaceship can get back by throwing something in the opposite direction − the heavier the thing he throws, and the faster he throws it, the sooner he will get back.

Another procedure that would encourage linking is to include questions in tests that check students' associations. These questions might be 'open-ended, such as 'Write five propositions that you associate with the experiment on the flocculation

Table 8.1 A task of matching experiments with propositions

Against each of the following experiments write the number(s) of the statements that you think are associated with them.

Experiments
   Reflection in a plane mirror
   Images formed by a converging lens
   Waves on a coil spring
   Refraction of waves
   Young's experiment
   Diffraction of light by a single slit

Statements
   1 Light travels at 3×10^8 ms^{-1} in a vacuum.
   2 The speed of light depends on the medium.
   3 Light behaves in some ways like a wave.
   4 Light interferes.
   5 Light diffracts.
   6 Pulses reflect from fixed points with change of phase.
   7 Pulses pass through each other unchanged.
   8 A node is formed where two waves meet out of phase.
   9 Light rays diverge from objects and images.
   10 Light passes through real images.
   11 $\dfrac{1}{v} - \dfrac{1}{u} = \dfrac{1}{f}$
   12 $\dfrac{\sin i}{\sin r} = \text{constant}$

---

of clay', or might require students to match a list of experiments with a (probably larger) list of propositions and intellectual skills (Table 8.1). Mackenzie (1979; Mackenzie and White 1982) devised a test of linking between episodes formed on an excursion and propositions, that could readily be applied to practical work (Table 8.2). The stem of the question refers to an experience that the teacher knows that the students have had, while of several plausible alternatives one is a proposition that the teacher had tried to get students to link with the experience. For the example in Table 8.2 that was alternative C.

As well as devising specific procedures to encourage linking of episodes from the laboratory with propositions and intellectual skills, attempts should be made to improve general learning styles, so that processing of experience, including linking, occurs as a matter of course. Unfortunately such processing is now the preserve of a small proportion, the superior students. Studies of learning styles (for example, Baird 1986; Baird and White 1982a; 1982b) indicate that Tasker's student who merely followed directions without thinking is not unusual. Studies also show that better styles can be encouraged, though this takes a long time (Baird 1986; Baird

Table 8.2 Example of a question testing linking between an experience and a proposition (from Mackenzie 1979)

In the saltmarsh zone of a mangrove coast, you find growing everywhere a short spindly, bushy, scratchy plant called samphire bush. You visit several mangrove swamps around Westernport Bay. You see that samphire bush is about the same height everywhere – up to your waist.
You notice, however, that in some places it is a lot easier to walk across the saltmarsh through the samphire bushes.

Which ONE of the following facts does this make you think of?

A   Mangrove swamps are generally quite muddy.

B   Spring tides occasionally wash the saltmarsh zone.

C   Plant density equals numbers of plants divided by area sampled.

D   The distance between high and low water mark is always vastly greater than tidal range.

E   None of these facts.

What else did you think of as you read the situation? (Write on answer sheet).

---

and Mitchell 1986; Paris *et al.* 1984). The main principle in training better styles is that the students must take on responsibility for their own learning.

Methods of teaching can be devised to promote that. Among these methods are discussions of phenomena, such as floating or the distribution of deciduous and evergreen plants, in which the teacher withholds his or her view for as long as possible; encouragement of students to ask questions in class (something that is much rarer than most teachers care to admit); giving students notes, then asking them to turn them into diagrams; and journals or 'thinking books' in which students correspond with their teacher about their work (see Fulwiler 1987, for some interesting examples).

It is necessary to consider general learning styles, for learning behaviour in the laboratory cannot be divorced from learning behaviour in other classes. Tasker's student, quoted at the beginning of this chapter, surely was not purposeless in the laboratory alone. A student who follows directions mindlessly in practical work is continuing a mindless pattern from other lessons. Similarly, one who thinks in other lessons will think and learn well in the laboratory. This leads to two final points. First, the discovery of poor learning behaviour in the laboratory should be appreciated not as a matter of concern for practical work alone, or even just for science teaching, but as a danger signal for all schooling. Second, although improvements can and should be made in the conduct of practical work, real progress in raising the quality of learning in the laboratory depends on a general lift in learning style. Therefore, those who are concerned to make laboratories useful centres of learning should attack on two related fronts – specific procedures for practical work, and overall learning style.

Success in these attacks might lead to a revision of Tasker's interview:

Interviewer: What have you decided it's about?

Pupil: I'm going to find out by spinning these wheat seeds how the field of acceleration affects the directions in which roots and shoots grow. Instead of going up and down, I think the shoots will go towards the centre of spin and the roots away from it. But they might go up and down a bit, because gravity is still there. And I'm wondering what a plant would grow like in a spaceship, without gravity. Tangled mess, I guess.

Regrettably, this revision is an invention, for no real instances of parallel quality are at hand. They should become common through application of the principles outlined here from information-processing and constructivist theories of learning.

PART IV

The reality of practical science

Factors affecting success in science investigations

Kok-Aun Toh

Introduction

In spite of the growing interest in practical work in school laboratories since the 1960s, research has provided little evidence as to what helps students be good at investigations. This chapter will specifically deal with research into an urgent, yet neglected, area – that concerning the factors affecting performance in investigations. Should such 'skills' represent a major intended outcome of the science curriculum, then it is imperative that pupils do well. This, in turn, would mean that there is a need to know what helps pupils perform well in investigations with a view to incorporating this into curriculum plans and translating them into actual practical work.

The purpose of this study is to investigate the factors affecting performance in investigations of 13-year-old students. More specifically the following questions will be dealt with. First, how do the factors of prior knowledge, attitude to school, attitude to science, and academic self-concept affect success in the performance in science investigations? Second, what is the contribution of explicit understanding towards success in performance in science investigations? And third, is tacit understanding adequate for the performing of science investigations?

Related literature

Prior knowledge

A pupil's prior knowledge is a factor likely to affect learning; this is best summed up by Ausubel (1968) when he said that the most important single factor influencing learning is what the learner already knows. Ausubel's model of learning has been elaborated upon and interpreted for science education by Novak (1978). Quite a number of studies have attempted to test the model for goodness of fit, these are described by West and Fensham (1974) in relation to prior knowledge.

Research studies have also confirmed prior knowledge as contributing to science achievement. Not surprisingly, Boulanger's (1981) meta-analysis of research gave a mean correlation of 0.46 between achievement in science and measures of prior knowledge.

Affective factors

Attention to affective variables seems to stem from the belief that they are as important as cognitive variables in influencing learning outcomes (Schibeci 1984). In support of this, Head (1985) argued that 'the ability to perform a task and a willingness to do so are necessary for success'. Of the two, the latter may be the more important since it determines the personal response of the individual to the learning situation.

The development of positive attitudes towards science has been regarded as a legitimate goal of science education and is well reflected by the inclusion of affective aims in most science curriculum projects. Bloom (1976) attributed as much as 25 per cent of the total variance in achievement to affective behaviours ($r = 0.50$), while Walberg (1970) considered motivation, self-concept and students' willingness to persevere intensively on learning tasks (all of which are affective) as significant predictors of learning outcomes.

Alvord (1972) sought to answer the question of whether students with a favourable attitude towards school achieved better. In his study of a sample drawn from the fourth, seventh and twelfth grades, he found low but significant correlations at all three grade levels. He felt there was a need to plan classroom instruction geared to focusing on both affective as well as cognitive goals.

Hansford and Hattie (1982), in their meta-analysis of research in this area, concluded that the overall correlation, r, between measures of self-concept and academic performance lies on average in the range of 0.21–0.26. A closer examination of the measures used in many of these studies shows that the more global the measure of self-concept, the lower the correlation with academic performance. On the other hand, with specific subject self-concept measures being used, correlations tend to be higher, $r \geqslant 0.40$. The study of Khor (1987) of specific subject self-concept and its relationship to academic achievement, for example, found correltations ranging between 0.38 and 0.51 for school subjects, including mathematics, English and science.

Tacit and explicit understanding

The literature is divided between the contribution that tacit understanding and explicit understanding can provide in the performance of laboratory investigations.

The case for tacit knowledge as a personalised knowledge that comes through experience has been lobbied for by Woolnough and Allsop (1985, p. 72) when they argue: 'We often pretend that focal, articulated knowledge is the highest aspiration and that having acquired this we use it to solve problems. In reality most problem solving is done directly through tacit knowledge, acquired through personal experience.'

Tacit knowledge has its genesis in the work of Polanyi (1958). He presented tacit knowledge as knowledge which we possess but which cannot be described or verbalised. Like cycling, it is acquired through personal experience and you cannot convey to someone, especially a non-cyclist, how it is that you cycle. Polanyi aptly described this as a 'knowing of more than you can tell'. It is likely, therefore, that tacit knowledge is not internalised as part of the superordinate structure of things, as would be the case for articulated knowledge. It is nevertheless there and serves to inform articulated knowledge – leading to full 'understanding'.

The knowing of more than you can tell is evident in the work of the Assessment of Performance Unit (APU) when it reported that pupils often controlled a variable (as evident from their checklist records of what actually took place) but did not reflect this in their write-up after the investigation. The pupils who were actually involved in the practical investigations could also perform higher-order tasks, such as control of multiple variables, but were unable to make them explicit in later interview/discussion sessions. This prompted Gott and Murphy (1987) to say that 'tacit understanding rather than explicit understanding of the concepts involved can prove adequate' for the performing of investigations.

In the context of the APU surveys, which were 'snapshots' of the prevailing scenario, the inability of students to make explicit what they actually did could be interpreted as being due either to the lack of appropriate language (words like control or vary), or to ideas of control or fair test not having been made explicit. This would mean that pupils were using their tacit understanding of these concepts to solve a given problem. They were knowing more than they could tell.

The question of providing explicit instruction on a particular aspect of investigative work was the focus for a number of studies (for example Rowell 1984 on the controlling of variables; Pouler and Wright 1980 on hypothesis generation). A detailed framework which reduces the whole of scientific activity to a composite of many small parts has also been attempted with some fervour (see, for example, Bryce *et al.* 1983). There is doubt, however, as to the wisdom of equating the whole investigative process with the practising of experimental skills (Woolnough 1989a; Solomon 1988). Unfortunately, a gap exists between translating these opinion-based doubts into research-based evidence. The provision of explicit instruction for the fundamental understanding of the nature of scientific investigations is still an area that is avoided by researchers. Such instruction easily extends from the basics of formulating testable hypotheses to the developing of a strategy to solve problems at hand, and the specific concerns given to the sensitivity of the available measuring instruments. These concepts are not easy to understand and the following remarks by Tamir (1989c, p. 62) show his conviction: 'explicit instructional efforts should be devoted to them . . . [as] there is enough empirical evidence . . . [to] indicate that most students will not learn . . . [them] "in passing" '.

The debate concerning the relative importance of tacit understanding and explicit understanding of the fundamentals of investigative work continues and what is needed is hard empirical evidence to substantiate one or the other.

The instruments

For the purpose of this study it was necessary to design or develop: a student questionnaire to provide a measure of the affective variables (attitude to school, attitude to science, academic self-concept); a prior knowledge test to provide a measure of influence that prior content knowledge would have in determining success in investigative work; and a suitable and accurate measure of student performance at investigations that can be used by pupils for reporting on their investigations.

The student questionnaire

The Student Questionnaire is an adaptation of the NFER Student Attitude Questionnaire (Skurnick and Jeffs 1970) and the NFER School Questionnaire (coded as SF.7). An initial 30 items were pilot-tested, following which some items were altered for the actual study. Factor analysis by principal components indicated a distinct three-factor solution (attitude to school, attitude to science, and academic self-concept) on 27 items of the modified version of the questionnaire. The Cronbach alpha-coefficient for the 27 items is a respectable 0.87, and the median alpha for the three subscales is 0.76 – showing good internal consistency.

The prior knowledge test

The test items were confined to content area relevant to the investigative tasks administered, so as to ensure content validity. This test (KR20 = 0.65) was reviewed by a ten-member expert team of science educators, then field-tested and subjected to item analysis, before being used for the study. In the main, the items were not of a recall nature. They were intended to probe the higher-order skills of understanding and application rather than that of the lower-order recall type. This view was taken because the area of study is that of scientific investigations, where students would be expected to exercise their command of higher-order skills rather than that of the recall type.

Measuring student performance

Student performance in laboratory work can be measured either by teacher observations or by the use of some form of report submitted by the student on completion of the project or investigation. The former suffers from pressure on teacher time. It is also uncertain how far such an assessment actually matches the pupils' overall performance unless the teacher is able to observe an investigation in its entirety on a one-to-one basis. The latter may be used if it can be shown that what is put on paper truly reflects the pupils' action. As a preliminary part of the research into the factors affecting a pupil's ability to perform practical investigations, it was therefore necessary to devise an instrument to measure the pupil's performance both reliably and validly. It was decided that a form of written assessment be developed following the APU's (DES 1985a, p. 33) finding that, in general, 'pupils asked to write an account of the investigation which they had just completed did so very accurately. Discounting vagaries of expression, pupil accounts of what they did matched very closely the supervisor's record of their actions.' To test the efficacy of different types of written report, three types of report sheet were designed, one open-ended (uncued), another specific-focused (fully cued), and a third broad-focused (partially cued).

The open-ended report sheet was designed to allow students to report on their investigations without attempting to influence what they have to say. The specific-focused report sheet had questions which were intended to elicit specific responses from students. Certain questions were generic and were similar across different investigative tasks. Others were task-specific and not repeated across tasks. This second type of reporting provides a more mechanistic form of reporting. The third report sheet adopts an intermediate style. It attempts to cater to the less articulate students by providing them with a report sheet which focuses on the broad areas corresponding to the different stages of progression through an investigation. A study of the literature (for example, DES 1981; OCEA 1987) has indicated that six broad stages can be identified: preliminary trials, planning, performing, communicating, interpreting, and feedback. The broad-focused nature of this third report sheet makes it applicable for the reporting of all investigations, yet provides some form of structure upon which students are able to make up a sufficiently coherent report. The details of validating this form of assessment have been reported elsewhere by Toh and Woolnough (1990).

These report sheets were tested for their accuracy in depicting what was actually carried out by comparing them to the supervisor's observational checklist – one that was arrived at on a one-to-one basis. Overall correlation coefficients of 0.70, 0.80 and 0.67 were obtained ($N = 61$) for the open-ended, broad-focused and specific-focused report sheets, respectively (see Table 9.1). This shows that the broad focused report sheet is the best, giving a correlation of 0.80 with the supervisor's score.

The study

A four-group design (see Table 9.2) was necessary to examine the effects of practice alone, practice plus instruction, and practice plus instruction plus pre-test. The

Table 9.1 Correlations between report sheets and supervisor's checklist

Investigative task	N	Pearson's r for report sheets		
		Open-ended (uncued)	Broad focus (partially cued)	Specific focus (fully cued)
A	23	0.78**	0.81**	0.48**
B	20	0.74**	0.84**	0.77**
C	18	0.54*	0.74**	0.70**
Overall	61	0.70**	0.80**	0.67**

Notes: * $p < 0.05$
       ** $p < 0.01$

Table 9.2 The four-group design

Group	Pre-test	Treatment	Post-test
1	Yes	IP	Yes
2	–	IP	Yes
3	–	P	Yes
4	–	C	Yes

Legend: IP – Instruction plus practice
        P – Practice
        C – Unrelated activities (Control)

question of inclusion of a pre-test (involving the actual conduct of hands-on investigation) was explored because of the influence it might have in sensitising subjects by providing them with the opportunity to practise.

There were essentially three experimental groups (groups 1, 2 and 3) and one control group (group 4). Group 1 was subjected to a pre-test comprising three short APU-type investigative tasks. Each task was to be carried out in 30 minutes, and students were given a further 20 minutes for reporting their investigation. This was followed in consecutive weeks with a sequence of practice–instruction–practice treatment using the learning cycle developed by Karplus (1977). During each of the practice sessions a single investigative task was provided for students to try their hand. A post-test similar to the pre-test was administered at the end of the sequence. The whole sequence covered a period of eight weeks.

Group 2 followed the same pattern throughout as group 1, but without the pre-test. Group 3 did the same thing as group 2 but without the instruction, primarily to study the efficacy of the practice component alone on performance in investigative tasks. During the time they were not provided with instruction they reverted to their scheduled class routines. Group 4 was given only the post-test in the final week in tandem with the other groups. During the whole duration prior to the post-test they continued with their normal lessons under their science teacher. These were activities unrelated to the performing of investigative work. All four groups had the same science teacher.

The instruction was provided through a decision-making package. The package involved the students in active and meaningful learning by including group discussion of problem situations to obtain consensus. This was achieved by involving the students in the generation of decisions required for each stage of the investigation. The decisions were cast in the form of questions which students had to respond to. For example, there were five decisions students had to make while 'planning' their investigations:

1 What do I have to measure?
2 What do I have to vary?
3 What do I have to control?
4 What are the things that make my results inaccurate?
5 What measurements need to be repeated?

A total of 15 decisions would have to be made and these decisions served to guide students through the different stages of the investigation. Particular care was exercised to ensure that these decisions were broad-based so as to be applicable to different investigations rather than fitting a specific one.

The sample

The sample, comprising 277 students (of which 170 were males and 107 females), was drawn from two urban schools selected from the mid-range of schools in Singapore. The two schools were considered as representative of a sizeable number of schools in the country. A total of four classes at the beginning of their eighth grade (13-year-olds) were selected from each school. Intact classes were used for the study as the curriculum and school set-up did not facilitate random placement into different treatment groups. The four classes in each school were fairly homogeneous because of centralised posting of students of similar ability, according to the nation-wide examinations administered to students at the end of the sixth grade, by the Ministry of Education. Additionally, the researcher administered the NFER-AH4 General Ability Test (Heim 1970), which incorporates verbal, numerical and spatial attributes, to ascertain whether the groups were comparable. The analysis of variance (ANOVA) on the performance in this General Ability Test showed no significant difference across the four groups in each school at the 0.05 significance level. Comparison across schools also showed that the schools were not significantly different at the 0.05 level.

Results and discussion

Effect of tacit and explicit understanding

The investigations were graded on each of the six stages, and their aggregate represents performance in laboratory investigations (POSTTEST). The outcomes

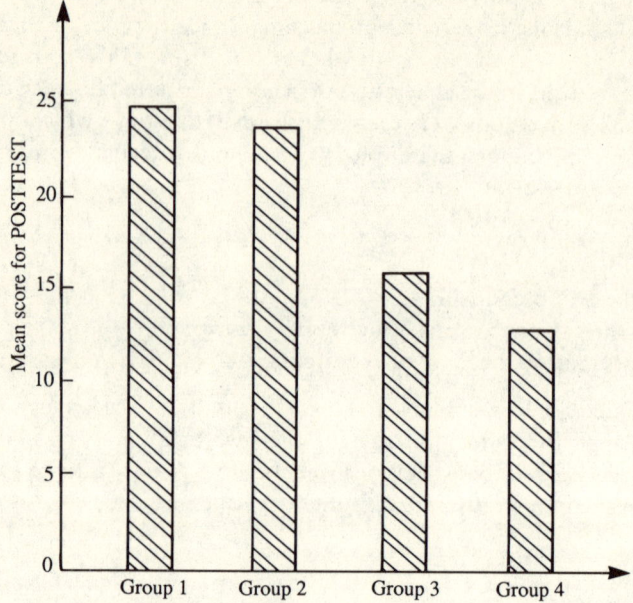

Figure 9.1 Performance in investigations (POSTTEST) by different treatment groups (maximum possible score = 48)

from both schools were very similar. Because of this similarity the results of the two schools were put together to give a single larger sample. Figure 9.1 shows the outcomes for each of the four groups.

The results, subjected to ANOVA, showed a significant difference between groups for both schools. The Scheffe procedure ($\alpha = 0.05$) that was employed to follow up the results indicated that, except for the difference in mean scores between groups 1 and 2, the differences for comparison between groups are significant. This would mean that either the pre-test did not sensitise the subjects by providing them with additional practice, or the combination of instruction and practice provided during the treatment overshadowed any practice effect afforded by the pre-test.

There is a big drop in performance as we move from groups 1 and 2 to groups 3 and 4 (see Figure 9.1). The instruction effect as shown by the difference in group means between both groups 1 and 2, on the one hand, and group 3, on the other, is significant at the 0.05 level (using the Scheffe procedure). This would imply that the provision of instruction for the conduct of investigative work provided to groups 1 and 2 had substantially improved their performance in laboratory investigations as compared to group 3. The latter, which can be dubbed as the 'tacit group', had all the experiences of group 2 but without any instruction on how to go about carrying out the investigation. Group 4 is the control group, and its difference from that of groups 1 and 2 is even more striking (see Figure 9.1). The combined effect that instruction and practice brought about shows the reinforcement that practice can provide to instruction.

The difference in group means between groups 3 and 4, though significant at the 0.05 level, is smaller. This shows that practice alone, without instruction (as in the case of group 3), will provide a certain measure of success in performing investigative work when compared to those who have none (as in the case of group 4). In symbolic terms all these differences can be represented as

Group: 1   >   2   ≫   3   ≫   4
                           ↑   ↑   ↑
Difference due to: pre-test instruction practice

where ≫ means that the difference is statistically significant at the 0.05 level, and > means that there was a difference but one that was not significant at the 0.05 level.

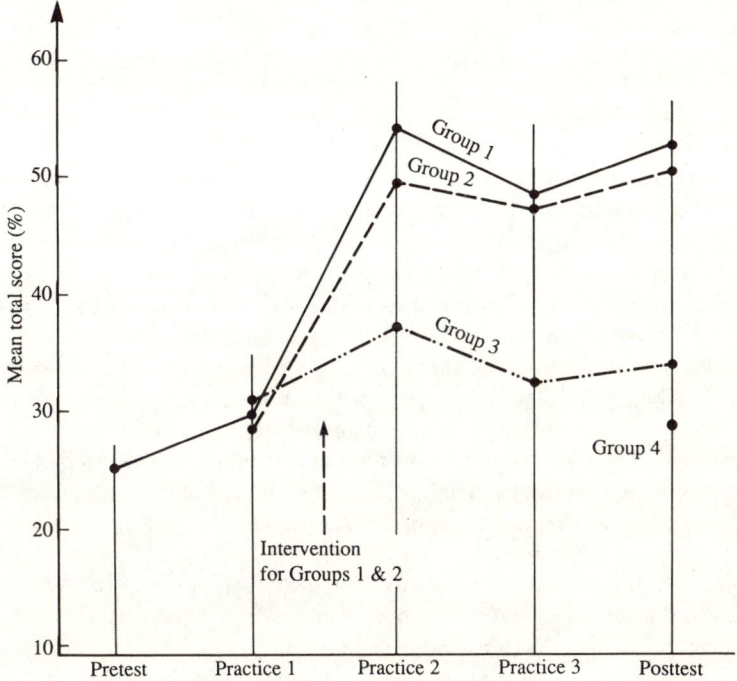

Figure 9.2 Progression of performance over duration of study
NB The lines joining the performance level at each stage provide for ease of comparison and should not be interpreted in any way to represent some linear progression from one point to another.

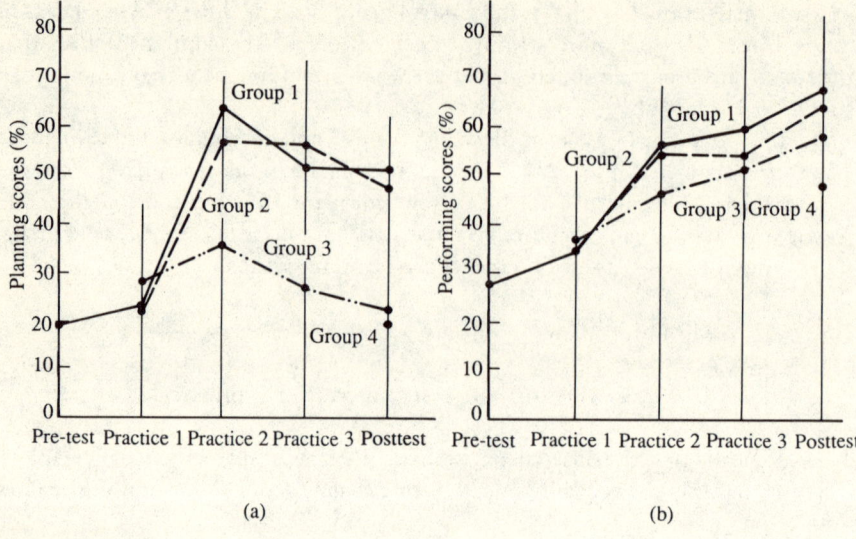

Figure 9.3. Planning (PLA) and performing (PER) scores

Additional insight can be obtained by considering the progression of pupil performance over the entire period of the study. Figure 9.2 shows the progression of performance as a whole, over the duration of the study. The overall pattern of 'ups and downs' is very similar for groups 1, 2 and 3, though that for group 3 is more subdued. Although they are more or less similar at the start (at Practice 1), subsequent intervention has resulted in widely differing levels of performance.

Figure 9.3 shows the progression of performance in two aspects, planning and performing. Figure 9.3a shows that pupils in group 3, the tacit group, did not perform as well in the 'planning' aspect of the investigations as those in groups 1 and 2, where the concepts involved for planning had been made explicit through instruction. This is to be expected, considering that pupils in group 3 had not been provided with the necessary language to express the concepts of planning such as control of variables, or fair test. One would therefore expect their 'performing' skills to be similarly depressed.

An examination of Figure 9.3b, however, provides evidence to the contrary. Pupils in the tacit group (group 3) are 'performing' much better than they are able to plan. Their performance comes close to that of groups 1 and 2. This is an indication of their being able to do more than they can tell. To put it another way, the pupils are using their tacit understanding to carry out the investigations. Although they are unable to make explicit their ideas in planning they are, nevertheless, able to perform.

Table 9.3 Intercorrelations among the predictor variables (ATTSCH, ATTSCI. ACADSC, PRKNOWL and SEX) and the criterion variable (POSTTEST) for groups 1 and 2 [$N = 135$]

VARIABLE	1	2	3	4	5
1 ATTSCH					
2 ATTSCI	0.53*				
3 ACADSC	0.54*	0.45*			
4 PRKNOWL	0.41*	0.47*	0.35*		
5 SEX	−0.09	0.14	0.14	0.13	
6 POSTTEST	0.66*	0.61*	0.64*	0.63*	−0.04

* Significant at the 0.01 level

Relationship between factors

Table 9.3 shows the correlation between success in performance in laboratory investigations (POSTTEST) and the following variables: attitude towards school (ATTSCH); attitude towards science (ATTSCI); academic self-concept (ACADSC); prior knowledge (PRKNOWL); and sex (SEX). The high correlations (r in the range 0.61–0.66) of all the variables (except SEX which is −0.04) with success in performance in laboratory investigations seems to bear out the equal importance of attitudinal as well as cognitive factors for success (Head 1985). Those with high attitudinal scores (that is, high scores for ATTSCH, ATTSCI and ACADSC) as well as those with high cognitive scores (high scores for PRKNOWL) are high achievers in terms of success in POSTTEST.

The ATTSCH variable has a correlation coefficient of 0.66 with POSTTEST, implying that 43 per cent of the variability in the POSTTEST is accounted for by this variable. This was also confirmed to be the single most important predictor for success in performance in laboratory investigations in multiple regression analysis. The data obtained do not indicate any significant relationship between SEX and POSTTEST.

Conclusion

The substantial difference in the performance of those provided with instruction compared to those without, would attest to the efficacy of the instruction provided. There are two implications for this: first, the teachability of the fundamentals of carrying out scientific investigations, to provide the explicit knowledge necessary to proceed with the investigation; and second, that such knowledge would be best served by explicit instruction rather than by mere practice. Undeniably the presence of practice is a vital element after instruction to provide the necessary reinforcement, and it is unthinkable to have instruction without practice for investigative work. But practice alone may not provide sufficient enlightening of some of the fundamental understanding of the nature of scientific evidence. These findings are in line with those of Tamir (1989c).

Notwithstanding the above, practice provides familiarity. And with familiarity comes confidence and the ability to think beyond the immediate. It was that the provision of repeated practice would bring out some of the requisites of scientific investigations, no matter how crude initially, which can then be refined later. In other words it was anticipated that the provision of practice would bring forth tacit understanding in the performing of investigative work.

There are two separate pieces of evidence for this: first, the significant difference ($\alpha = 0.05$) in the overall mean scores of groups 3 and 4; and second, the relative success of pupils in the 'performing' aspect of an investigation when they are unable to carry out the 'planning' aspect of it. The latter is a reflection of their 'knowing of more than they can tell', thus corroborating the findings of APU (DES 1987a).

The study thus points to the benefits that can accrue through tacit understanding (by use of practice) but identifies deficiencies in purely providing tacit knowledge when it involves more abstract concepts. To internalise the more abstract concepts explicit understanding (by use of instruction) may be the most efficient way.

The evidence from this study also confirms that both affective and cognitive factors play an important part in success in investigations, in corroboration with documented studies. One must qualify this, however, by saying that the values quoted for other studies may be for science achievement in general and not specifically for hands-on scientific investigations as in this study – they are, therefore, not directly comparable.

To the extent that the affective and cognitive variables are stongly correlated with success in the performing of science investigations, the four factors explored collectively account for more than 68 per cent of the variability (by multiple regression analysis) in performance of scientific investigations. It may also be worthwhile to point out an important finding which may seem to be an intuitively accepted link – that a good attitude to school might contribute to success in general. There is much interest here but so little in the form of documented evidence especially for open-ended investigative work. The evidence adduced from this study is that attitude to school accounts for 43 per cent of the variability in success in performance in investigative work, and that collectively the affective factors explored account for 53 per cent. This points to the importance of affective factors in contributing to success in open-ended experiments. It is, therefore, not surprising to see such outcomes expressly stated in most curriculum statements of aims, though the same cannot be said for those of assessment.

CHAPTER 10

School laboratory life

Joan Solomon

In this chapter the focus of attention will not be on the achievement or assessment of learning during practical work, in the more usual sense of how much content or process has been assimilated. Instead, it will give accounts of laboratory behaviour, private and public, individual and group, including and excluding the teacher.

In the usual school vocabulary 'behaviour' implies *bad behaviour* which detracts from learning and so should be eradicated. In other context behaviour may be either good or bad but is thought to be easily distinguished from learning. The perspective in this chapter is topsy-turvy. It will seem to ignore all the teaching and learning of science except in so far as it forms part of the backdrop to laboratory experience. Yet if the accounts of how children and teachers interact with each other among all the words and apparatus which clutter laboratory existence are vivid and accurate enough, they will be validated simply by recognition from those who have been a part of such scenery. If the accounts trigger recollection of the minutiae of doing science in the school laboratory, then, in the manner of all recollection, it will recall the whole scene together with the feelings of surprise, or boredom, or excitement. Almost incidently, we may also come to understand some of the peculiarities of learning through experiment, because this too is a part of the general ambiance of 'doing' school science.

Collecting material for this scene painting claims to be research in the tradition of insider ethnomethodology. It is based on reports given by science teachers themselves. Some have argued that any valid study of a 'normal' working laboratory should exclude outsiders, not only in order to maintain the status quo which is being studied, but also because interpretation is an inalienable part of observation and it is the inside actor's interpretation which will be much the most insightful. That school of research would lie at the opposite extreme to more positivist schedules of interactional analysis. It would even have to reject *post-hoc* interview studies of students and teachers about their experiences in the laboratory because these would be open to the claim of being fabricated after the event.

In this research tradition the teacher is just one more actor on the scene. Of course, teachers are trying to make sense of what goes on – as we all do throughout life – but theirs will be only one sense of the total activity. Their comments are not presented as objectively 'true' accounts of how things 'really' are; rather they are just a part of some total account of all the accounts which might have been given by the actors on the scene.

Making arrangements in a strange space: listening to talk

Young children have a very vivid sense of place. Indeed, it is so memorable that adults may still site their dream lives in the locations that they knew many years before in chidhood. So it happens that if primary or young secondary school children are asked about the sport or science that they did in primary school they may even deny that it happened at all: 'No, because we had no proper football pitch there' or 'In primary school we had no bunsen burners. We did no science then.' By the same token they look forward to science in the 'big school' because it will take place in that strange place, the laboratory, with all the paraphernalia of science fiction – bubbling flasks and frothing test-tubes. Some interesting work in progress shows that first-year pupils in the secondary school will still draw pictures of a laboratory full of these exciting stereotypic experiments, even when their own year's worth of school science experiments has been exclusively with forcemeters and circuit-boards.

What the primary school pupils probably anticipated far less clearly was the complexity of the group tasks that they would be given. Even when the teacher very carefully demonstrates the fitting together of each piece of apparatus and from where it should be collected, a great deal of negotiation must take place within the group doing practical work together in order to arrange who shall fetch which piece, who shall take first turn at the more coveted tasks and, more difficult still, who has the least popular task of clearing up and putting away.

Every science teacher has heard fragments of this kind of squabbling activity but many will have failed to see it as a part of the whole stream of group talk which serves to make practical collaboration possible. In a listening and observing study of a low-ability second-year class doing practical work, Wallace (1986) patiently recorded all the snippets of talk and activity that she could hear or see. Nothing was superfluous: she wrote

It looks as if children need to chat about anything as well as their work. Similarly play, another informal activity, seems to give opportunity for developing understanding. Although this must be commonsense knowledge to many teachers, there are some who feel guilty about allowing these informal activities . . . Chatting is discouraged, and, so too is 'messing about', the derogatory term used for play (p. 72).

In Wallace's study the pupils worked in special science pairs and the observer sat inconspicuously among them. Having arrived from other lessons, the children came together with opening comments, about football, or a sore finger, rather like birds squawking and fluffing out their feathers as they settled down in a nest. During the lesson they taught each other by repeating or translating what they thought the teacher had said, they helped, they argued, they agreed, they consoled, they described and they gossiped. Wallace listed the categories of talk under six headings:

1 Negotiating doing (for example, arranging the collection of apparatus and turn-taking in the experiment).
2 Removing tension (for example when disappointments or near-quarrels have occurred.)
3 Giving help and tutoring (often in the teacher's words.)
4 Non-task talk for greeting or 'stroking'.
5 Negotiating perception (for example, agreeing upon what colours they are seeing or what the measurements come to.)
6 Suggesting a meaning or explanation for what has happened in an experiment.

Authoritative documents about how to encourage good investigations have always tended to focus on category (6) and to ignore the rest. Yet Wallace reports that this was by far the rarest kind of talk. At several points in her commentary she emphasises that *all* these categories of talk are essential for the social activity of doing practical work. Without category (1) the children would continually bump into each other, without (2) they might fight and without (4) it is almost impossible for any two individuals to enter into close and active proximity.

Category (5) is interesting. Perception may be thought to be a private affair taking place in the neural connections between the retina and the brain, and yet urgent cries to others of 'Look, look!' are commonplace in the laboratory. Wallace tells of a small boy who is unsuccessful in getting the attention of his friend to watch and corroborate what he has just seen, and calls instead for the teacher, as a kind of extension of the group, to fulfil the same function. The sight of a small girl with a frothing test-tube making her way across a busy laboratory to 'show Miss' is a common, sometimes comic, and potentially hazardous event; but the purpose is entirely serious. New effects are difficult to perceive, even to see, and certainly to relate to a scheme of understanding. Without social reinforcement none of us knows quite how to 'believe our own eyes'.

Doing teaching with the second year

From the teacher's perspective only some of these activities are noticeable and the emphasis is bound to be different from that of the observer or the pupil. What follows are notes, almost completely unembellished, which were written two months into the school year at the end of a second year's double period of science by their class teacher.

Right at the beginning of the lesson Martin brought up his special ironstone to show me. While the technician got out the power packs at least ten of the children clustered round me with cries of 'Look at my homework, isn't it good?' Mandy made excuses for not having done it, and Katy fussed as usual. (She has now deserted Christine and is working with Jane.)

I showed them how to get copper out of copper sulphate by electrolysis. No sooner had they begun than Martin touched the electrodes together at 12 volts and tripped the switch. I was cross – I had said 'only 5 volts', and also 'Don't let them touch'. The class was subdued for a while by my anger.

The rest of the experiment went well. Trudie and Nicola each came up to bring me to their experiment to admire their work. Yvonne showed me how she had electroplated a coin, explaining it was her idea – Nicola argued that it was hers!

Then several came to borrow coins from me. Alice asked if she could put her coin in her book as an exhibit. Later she came up again to show me what she had done and tell me not to lose the coin. I sent her to the technician to get some Sellotape to fix it.

Matthew was fussing a lot, showing off. Stephen and Graham got it wrong, I showed them that they had put the coin on the positive terminal. Graham blushed and put it right. They went on for far too long – later he showed me his coin in triumph.

Peter (he's bright!) showed me his book where he had written 'the copper is drawn like a magnet'. I suggested that he think out his theory so he left the group and worked by himself on the side bench. Later he showed me his book again. He had written 'there is no connection between the + and the − so the copper sulphate joins the circuit'. I spoke to him about attraction between + and −. He suddenly understood, went back and wrote 'If it is attracted to the negative it must be (+) positive'. I was pleased.   Graham and Matthew talk with me at the end of the lesson. Graham says his brother is 'a scientist' and that he has electroplated all his father's tools. As he walked back to the main school with me he said he wasn't going to be a scientist but a farmer like his Dad. I said he needed to know science for fertilisers; he said he was going to keep animals. Talked about injections. He said calves needed sleeping pills otherwise they stayed awake until they dropped. He was very confident and cheerful.

Next lesson Peter sat alone again on the side bench. He came up to remind me about it: 'You remember I didn't do the last part because I was writing?' I had to get him to join Martin (his group) twice. He was silent and angry; they didn't finish.

Much of that report shows the teacher's perception of herself as the locus of control, approval, rebuke and encouragement. Almost every comment confided to the notebook is judgemental, probing the children's action and comments for correct behaviour and correct reasoning. Affective chat about homework and argument about who first thought of using a coin go almost without comment. Peter may be encouraged, but must not separate himself from the rest. Only when they leave the laboratory can a pupil like Graham take control, reject the teacher's idea about knowing science in order to understand about fertilisers, and usurp her function by giving out information of his own about calves. Both of them, teacher and pupil, seem quite comfortable with this change in role. 'Doing science lessons' ends at the laboratory door.

The pupils try continuously to achieve a good reputation in the eyes of their teacher. Most of the ones she notices have moved across the laboratory to be near her so as to show off about their homework, or to fetch her to come back with them and admire the work they have done. Necessarily, we hear much less about those who do not jostle for approval and reputation.

Even the positions in which the pupils sit are related to the approval which they seek from the teacher. Peter, who longs to be acknowledged as different and an intellectual in the eyes of the teacher, sits in the middle of the front row in order to be noticed – if Martin will allow him to do so. He seems not to be popular with his peers. Is this why he seeks reputation from the teacher, or is his unpopularity a direct result of this effort to be appreciated? Similarly Nicola, who has privately asked 'Miss, do you think I could be a scientist?' also sits on the central aisle but further back and slightly separated from the rest of her group.

The 'fussers' – Mandy, Kristine and Katy – sit on the side benches, but in front. In this way they can face away from the teacher and chat to each other while they do their experiments. Here, too, they can create a stage on which to act out their little dramas to be watched by the teacher!

How to be a group

One group of these 12–13-year-old children changes very noticeably in its behaviour during the first six months of this year. It consists of four girls. At the beginning Karen comes to the front to talk about the experiments. She wants to speak about whether the 'fizz' from chalk and acid is the same as air, and offers to bring her chemistry set from home to show the teacher. Mary also comes forward to talk about the limestone she saw on holiday. The new girl in the group, Anna, waits until the teacher shows she knows her name by addressing a question to her in class, then she comes forward, too. The fourth girl, Sheila, keeps her distance.

For two months all, except Karen, stay immovably in their places alongside Sheila. Karen sits on the central aisle, fetches the apparatus, and still talks occasionally to the teacher, who praises her practical initiatives. Now Mary and Anna both blush (with irritation?) when asked a question directly and so are forced to speak to the teacher in class. Their written work is good but gradually they are adopting a group attitude of 'our experiments always go wrong' which excuses their slothful practical work. Karen is torn by this; she still wants to experiment in order to preserve her own reputation, but is eager to remain within the bounds of the permitted group behaviour. When the teacher approaches it is Karen who speaks to her, although by now rather uncomfortably: the others draw back and seem to feel 'invaded' unless they have themselves called for the visit.

In March a technological exercise gives this group a new opportunity to express its preferred behaviour. The group has been challenged to make a water turbine from a cork stuck with blades cut from a yoghurt carton, and to use it to raise a weight. The lesson becomes noisy and busy, except for this girls' group. After some half-hearted work they manage to make so poor a machine (with short and misplaced blades) that it can only raise the weight a short distance. Karen calls the teacher over to point out that the weight will rise only while it is under water. Seeing that the teacher is eager to speak about this apparent loss of weight, while the rest of the group are irritably demonstrating their lack of interest, Karen stops the teacher by saying 'Our's was a total failure!'. The others agree readily – with some contempt for a group of small boys next to them who are pouring the water straight down on the spindle of the turbine and so getting no movement at all.

Following a suggestion by the teacher the girls observe the efforts of the other groups and then write down notes on 'How to make an ideal turbine'. This enables them to reinforce group solidarity by doing no practical work at all of their own, and yet achieving superiority through criticism and writing, both of which they do well. They pause by John and Peter, whom they like. The boys' turbine is spinning furiously with a great spray of water. They smile and repeat that their own machine was 'useless' with apparent satisfaction. Peter is happy to explain, with all the technical detail he can manage (and half an eye to the teacher) how he and John have designed and built their turbine. The two groups have thus successfully negotiated their different characters and reputations.

The psychology of adolescent groups

Different strands of educational research need to be pulled together to understand the behaviour of normal adolescent groups. There are books on ethnography inside schools which never mention adolescence, and books on adolescence which are entirely concerned with self-image. There are whole libraries of works on deviant behaviour of individuals or groups which concentrate on maladjustment during adolescence. All of this is troublesome for a teacher or researcher who believes that group adolescent behaviour shows special and sometimes difficult features, which are none the less both normal and rational.

The substantial body of research on social development during primary schooling shows a steady increase in communication between peers. Indeed, the claim is often made that social learning is the single greatest achievement of primary education. By secondary school age the adolescent's need to feel 'grown up' leads to an active suspicion of most adult advice and knowledge, and to even stronger links with peers. 'As individuals become increasingly independent of their families, they depend increasingly upon friendships (as well as partnerships) to provide emotional support and to serve as testing grounds for new values' (Douvan and Adelson 1966). The provision of emotional support involves rules which can be relied upon and roles in which each may feel at home, except for those whose unenviable role it is to be ridiculed. Within the school science laboratory special rules and roles are forged which include all permitted kinds of action, ranging from how to talk to the teacher to who should light the bunsen burner.

This social perspective stems from the work of Erwin Goffman and has been extended by Rom Harre, who applied the same line of thinking to the space that children occupy in their work and play. This has been used in a recent study of behaviour in science (Solomon, 1989):

> All the pupils' actions can be seen as 'social acts' by considering the meaning they may have for the pupil in the context of all the others present. In the corridor or out of school pupils may speak to the teachers without the same constraints, but while carrying out an experiment talking to teacher can be an abrogation of group autonomy – unless the task is complete and the group demands praise. A friend from a neighbouring group may become an unwelcome spy who is listening in on their plans for investigations.

These social acts build up performances from which the 'moral career' of a pupil develops. This is the *reputation the pupils have*, or wish to have, in the eyes of others – peers or teachers. One may wish to be thought good at planning experiments, or good at drawing or 'writing up'. For others it becomes just as important to be 'no good at science'.

Social acts have meaning for classroom groups which have their own group honour and reputation. The importance of the groups, and their conceptions of honour, vary with both the age of the pupils and their gender. For younger children the teacher is a stand-in parent to be included in all the marvellous results they hope to get. According to some psychologists, the increasing demands for loyalty, conformity and intimacy that are made of friends are a particular feature of girls' groups. Because of the masculine image of practical science, a reputation for *being good at it* is more important to those boys who have no wish to run counter to stereotypic expectations. These are the ones that Head (1985) speaks of as having foreclosed in their early adolescent phase without sufficient self-examination.

All space within a familiar class room has social significance and meaning. Space is task-related and entry is governed by conventions. A teacher (or another pupil) standing too near a group to watch them at work is resented: the pupils may literally squirm with unease until the teacher moves away. Only the command 'Come here, Miss' permits entry. Getting these groups to move in order to watch a video, or to look more carefully at a specimen on the front table, can be quite difficult. Typically these space-centred older groups may refuse to shift, supporting and encouraging each other in their immobility with cries of 'We can see all right from here'. (Younger pupils would have fallen over each other in their haste to reach the teacher's bench first!)

Group learning

Such descriptive explanation of adolescent behaviour may be of some comfort to a struggling novice teacher, but it needs to illuminate learning processes if it is to have direct educational value. It warns us that listening to teacher instead of gossiping with close group members may run counter to the inclinations of these adolescents in at least two senses. Not only the existence of rules for action within group space, but also the wish to be independent of adults, suggests we look to other strategies for learning from group practical.

But almost all enjoy being successful. Although some pretend to a reputation of 'our experiments never work!', success swiftly belies this contrary stance which has usually been adopted only to save face. They also enjoy adapting the experiment by going beyond what the teacher appears to intend. This is particularly satisfying to groups who perceive the intolerable domination of adults – which they are trying so hard to escape – even in the teacher's instructions and worksheets.

Because group reputation is a public affair the group's perception of success is easy to gauge. Not while the experiment or its extension is being tried out, but as soon as the outcome is thought by the group to be reputation-enhancing, the teacher or the neighbouring group will be summoned to admire. The act of making the result public is a sure sign that the group has staked its reputation upon it.

In the teacher's diary from which quotations have already been taken there are several examples of this kind of triumphant group initiative:

Some 13 year old boys loudly demonstrating their energy machine which is powered by a weight falling, not from the edge of the bench as was written on their worksheet, but from the top of a high and wobbling cupboard.

A group of 14 year olds making a small bonfire of burning wooden spills to reduce lead oxide, instead of just mixing it with charcoal.

An image being projected on the far wall of the laboratory, and on the ceiling, with consequent huge magnification, instead of using the small screen provided. This involved pulling down the blinds of the laboratory, and some subsequent quarrelling and disruption of the other groups.

A group of low ability 16 year olds rapidly boiling a test tube of water by using three burning peanuts all at the same time, instead of just one.

Challenge to some 16 year old groups to find out and draw a light ray being coloured by internal reflection resulted in different group diagrams (paper sheets sometimes joined together to make a poster), elaborately and rather eccentrically drawn, being brought up to the front to be 'shown off'.

The individual and the group

The persepective of the previous section cannot obliterate all the personal learning features. There may be real difficulty in moving from one theoretical stance to another during theoretical research – 'theory-laden observation' has become a philosophical cliché, but the conflict is not severe in the classroom. There is a substantial body of research on the individual's influence within groups. We may also expect some of the same factors which enhance group learning to be present in more solitary activities.

Within-group tutoring of one pupil by another is a common and welcome occurrence. Research in mathematics has shown that it benefits both the giver and the receiver: only those who take no part in the discussion suffer a little in their rate of learning by losing time and concentration. Even when the tutoring is not intended, talk between group members conveys meanings and even implicit theories. A small-scale and unpublished piece of research with second year pupils who were carrying out elementary experiments on electricity with circuit boards showed that the pupils' models of current flow tended to converge to a group mean during the series of worksheet experiments. Since the agreed model was often not the correct one being taught by the worksheets it must have been the result of pupil talk. Perhaps, as each group spoke about the connections being made and the results observed, meanings about how the current flowed and lit the bulbs were formulated and exchanged. The work of Moscovici (1976) and others suggests, not surprisingly, that it is the view of the most insistent member which is the most likely to prevail within the group.

The following extract is from the work of the Children's Learning In Science (CLIS) project (Wightman *et al.* 1986, p. 295) and nicely illustrates the uncertain balance of persuasion. After a series of simple experiences and lessons spent talking about them in groups one pupil sums it up:

P: A couple of people changed my mind a bit about the shape of atoms and stuff like that . . . because I thought that perhaps they'd be all round but later – well perhaps they could be square, rectangular, triangular whatever.

I: Well what do you think now?

P: I think they are round.

I: You didn't convince them that they were round?

P: No.

I: Did you try?

P: Yea I tried to but they wouldn't take any notice of me.

This pupil and several others interviewed in this project expressed their preference for doing practical over the long discussions of 'theory' that this action-research necessitated. The researchers commented that a single dissonant practical result – like that of the girl who found no protein in the soil but believed that the experiment had 'gone wrong' – did not easily convince their pupils. However, they nowhere suggested that practical work might require just as long a period of time as did discussion to be effective in changing opinion or in building up new models. We are left with the impression that talk, and not action, is the key to conviction.

Experimenting on one's own

Even in the solitary extreme, when lone children work by themselves, some of the factors we have observed in group activities seem to be applicable. In particular, the need to make adaptations or extentions to the givens of the apparatus or the instructions becomes as important to the individual as to the group. Once again the child who has found this new path will seek an audience to demonstrate the effect. In this way he/she becomes an 'expert'. Reputation in the eyes of others is at least a part of the triumph.

But close observation of a child alone with scientific apparatus shows clearly the extended time-scale that is needed for imaginative explanation. In school science clubs or in interactive science centres, it is commonplace to find a child going through the same routine again and again. Each time that the identical hoped-for, and yet surprising, result follows on from the well-learnt preliminaries the child seems pleased and satisfied. A teenager in school may beg for the same computer interface and produce virtually the same plot again and again. A younger child at a museum display may dip both hands into a vibrating dish of sand to feel the liquidity of it and watch buried celluloid ducks bob up to the surface, time after time. Their abstracted faces give few clues to the interior purposes of this repetitive play.

Hodgkin (1985) documents this stage in practical learning. Either 'playing' or 'practice' are essential preliminaries, in his thinking, to the later exploratory phase of learning. When the routine is finally achieved to that point of certain expectation which means that successful internal mental modelling has taken place, the equipment becomes no longer a plaything but an 'explore-with'.

Michael Polanyi wrote of a similar effect which comes about as a result of the craftsman's long practice with his/her tools. He called it 'in-dwelling'. This is the familiar feeling that all car drivers have about the vulnerable surface of the chassis during a 'close squeeze'. Even their terminology suggests that it could be their own body which was about to be torn or dented. Others refer to the equipment or tool which is so deeply embedded in the personal probing process as a 'cognitive object'. Whatever term is used, the outcome is mental modelling so secure that both prediction and personal control can replace the original delighted surprise. In a cognitive sense the experiment and its results are now *owned* by the child.

Ownership is highly important to all learning, be it solitary or group. It is achieved by internal action which may be partially triggered by discussion with others, or by following practical instructions, or by taking notes from the teacher, or by none of these. The brain cannot have learning stamped upon it as one might brand a sheep. The student needs time to reach out and enfold the new ideas, squeezing them into a shape which will fit his/her own peculiar mode of thinking. This may not happen until the following morning, by which time sleep has brought its mysterious aid. Or it may need some serendipitous event of everyday life to provide an analogy for its operation.

However mysterious this achievement of internal modelling, we can often recognise its arrival during practical work. When individual children at play, or a group of students in the school laboratory, take over the whole process, tuning it to suit some new purpose of their own, we may be sure that playing has moved over into exploration. They have learnt enough from the activity to own it; they can stand confidently upon it, as on a platform. It extends their reach by allowing them to use their personal analogue, their mental model, to predict what might happen in another situation so that they can explore further into the uncharted territory of new experimental investigation.

CHAPTER 11

Gender differences in pupils' reactions to practical work

Patricia Murphy

Introduction

The low number of girls studying physical science after the age of 13 has been the subject of world-wide concern since the 1970s. The change of status of science from an optional to a compulsory subject in the curriculum of many countries is an attempt to alter this situation. This can only be effective, however, if the reasons for girls' avoidance of science can be identified and counteracted. Much is known about the differences in attitudes, experiences and achievements of girls and boys as they relate to science. What currently concerns educators is the way these factors interrelate to create gender differences.

Recent curriculum innovations in science typically include an increased emphasis on 'active' learning, to which this book testifies. To ensure that such innovations are beneficial to all pupils it is important to consider how gender differences operate in the practical context. Unfortunately, there is a confusion of messages coming from the literature about girls' and boys' attitudes to, and performance on, practical activities. Omerod (1981), for example, found that 'liking of practical work' was a significant discriminator for boys in all three science subject choices, in that it was an added incentive to study science. This was not true for girls. Yet recent studies looking at classroom intervention strategies have recommended practical work, or 'active' work, as a way of combating gender differences (see, for example, Hildebrand; 1989). Some national science surveys have shown large differences in favour of boys on practical tests (Kelly 1981), others either no differences or a trend in favour of girls (Department of Education and Science 1988b). Girls' lack of confidence in practical contexts and fear of practical equipment has also been commented on in several studies.

To understand these apparent contradictions it is necessary first to explore the nature of gender differences and the factors that determine them. This chapter looks at some of the relevant international research in the area and examines the findings in the light of the results of science surveys carried out in the UK.

Gender differences in achievement – the international scene

Outside the UK the well-known survey programmes of educational achievement are those of the International Association for the Evaluation of Educational Achievement (IEA) and the National Assessment of Educational Progress (NAEP) in the United States. The first wave of surveys for these programmes took place in the late 1960s and early 1970s. The questions used were almost all multiple-choice and the majority were strongly content-based. The UK programme of the Assessment of Performance Unit (APU) started later (1980–4) and used quite different measures of achievement. The science surveys included: three practical tests, measuring skills, observation skills and practical investigations; tests of science content; and other process skill measures (Murphy and Gott 1984).

The IEA survey found that on average boys scored considerably better than girls in science achievement tests in all 19 countries surveyed. A study by Kelly (1981) explored this finding further. The study focused on the 14-year-olds in the IEA sample from 14 developed countries. The IEA science tests were written tests which included measures of chemistry, biology and physics content; laboratory practice, that is, knowledge of apparatus and experimental procedures; and attitudes to science. The results again showed boys ahead of girls in every branch of science. The largest differences were for physics and the practical sub-test. The smallest difference was for biology, with chemistry intermediary. Girls in some countries did achieve higher scores than boys in others. However, the nature and magnitude of the gender differences were consistent across the various countries. The NAEP (1978b) programme in the USA similarly found boys ahead of girls at ages 9, 13 and 17 on tests of physical science content, and only small differences in favour of boys in biology.

The cross-cultural uniformity of gender differences in science achievement suggests that cultural factors alone cannot account for them. Kelly (1981) identified the 'masculine' image of science, common to all countries, as a contributory factor. She argued that as girls and boys have learnt to respond to gender appropriate situations, then a 'masculine' science will alienate girls and discourage their engagement with it. Kelly's study also showed that internationally boys had a greater liking for, and interest in, science. There were considerable variations between countries, indicating to Kelly that cultural expectations did influence attitudes, particularly girls'. She did note, however, that boys achieved better in science than girls with equally favourable attitudes.

The UK APU science surveys of 11-, 13- and 15-year-olds included tests of both the skills and content of science. Only a small number of the test questions used were multiple-choice. The results of the surveys are given in Table 11.1, along with

Table 11.1 Some international survey results

APU test	Results from APU	Results of other surveys		
		IEA	NAEP	BCSS
Use of graphs tables and charts	$B_{15} > G_{15}$			$B = G$
Use of apparatus and measuring instruments	$B_{15} > G_{15}$	$B > G$		$B > G$
Observation	$G > B$			
Interpretation	$B_{13,15} > G_{13,15}$			$B = G$
Application of:				
Biology concepts	$B = G$	$B > G$	$B > G$	$B = G$
Physics concepts	$B > G$	$B > G$	$B > G$	$B > G$
Chemistry concepts	$B_{15} > G_{15}$	$B > G$	$B > G$	$B = G$
Planning Investigations	$B = G$	$B > G$		$B = G$
Performing Investigations	$B = G$			

B denotes boys' performance, G that of girls; B_x denotes the performance of boys aged x. The symbol '>' should be read as 'better than'.

the related information from the IEA and NAEP surveys already discussed. In some cases there is a suffix to indicate the age at which the gender difference emerges. Where there is no suffix this indicates that the difference or similarity in scores occurred across the ages tested. The APU results demonstrate that gender differences in favour of boys increase as pupils progress through school.

Table 11.1 also refers to the results of another regional survey, the British Columbia Science Surveys (BCSS) (Hobbs *et al.* 1979). This is included because, like the APU surveys, pupils' achievements on both the skills and content of science are assessed. However, like the IEA and NAEP surveys, the tests were in written not practical mode and the questions were largely multiple-choice. The British Columbia results show boys' superiority to be restricted to tests of physics content and of measurement skills.

The IEA results found that boys continued to outperform girls even when their curriculum backgrounds were similar. This was not the case in the APU surveys. When pupils with the *same* curriculum backgrounds are compared all performance gaps at age 15, except those for applying physics concepts and the practical test-making and interpreting observations, disappear. The results show that even when able girls continue to study physics they do not achieve the same level as boys as a group. The gap in physics achievement between girls and boys established at age 11 increases with age (Johnson and Murphy 1986), a finding also reported in the other survey programmes.

The 1984 APU survey looks in some detail at pupils' interests and attitudes to science. The results show a polarisation in the interests of boys and girls across the ages. Girls' interests lie in biological and medical applications, boys' in physics and technological applications. The same polarisation is evident in the pastimes and hobbies reported by the pupils. Girls also saw science as having little relevance to the jobs they might choose, unlike boys. Indeed, the proportion of girls selecting a job drops if they perceive a high science content in it. At age 15 a markedly higher proportion of girls than boys describe physics and chemistry as difficult. These findings replicate the IEA survey results.

At the time of the last APU surveys the second round of IEA testing was being carried out (1983–4). Some of the results for different countries have been published which provide further information about the nature of gender differences in science. For example, the second international science study (SISS) for Japan and the USA (Humrich 1987) again showed 14-year-old boys ahead of girls for each science subject. However, Japanese girls outperformed boys and girls in the USA in physics whereas Japanese boys achieved lower biology scores than boys in the USA. Japanese scores were lower than US scores on tests of knowledge but substantially higher on tests of comprehension and application. The differences in levels of achievement for girls and boys between the two countries appear to reflect the different curricular emphases. For example, in Japan there is an early focus on the development of reasoning skills. Mathematics, which is particularly influential in physics learning, is a highly valued subject for all pupils. Biology, on the other hand, is accorded relatively low priority. Clearly the educational objectives of a culture can influence gender effects.

The SISS results for Israel (Tamir 1989b) showed a similar pattern of differences to the first IEA study. However, by age 17 the gender differences in biology and chemistry had disappeared for science majors and the difference in practical laboratory skills evident at age 14 was not present at age 10 or 17. The study found that girls specialising in physics did less well than boys in all science areas. Whereas female chemists and biologists only did less well in physics. It is evident from these results that girls' learning in physics is an international problem.

National survey results have consistently shown boys ahead of girls on physics tests. Recent research in Thailand, however, provides an exception to these findings (Harding *et al.* 1988)[1]. Seven sets of tests were used in the study: three practical tests (of manipulative skills, problem solving, and observational skills) and four paper-and-pencil tests (on the links between practical work and scientific knowledge; sources of evidence; scientific knowledge; and scientific attitudes). The results show girls at age 16–18 performing at least as well as boys in physics and better than boys in chemistry. In laboratory tests girls outperformed boys in both physics and chemistry. The researchers posit several reasons for these uncharacteristic results, among them that science in Thailand is compulsory; the teaching approach is practical; chemical tasks have a 'feminine' image, as do some of the practical physics tasks; there are no differential expectations of pupils; and females participate in all levels and fields of employment.

It appears possible to alter girls' underachievement in physics by reconsidering the cultural expectations of girls and boys and how these are reflected in the organisation and values underpinning the school curriculum. There are indications from each of the national surveys that in certain circumstances tasks in biology and chemistry can favour either boys or girls. To understand these results it is necessary to examine the effects of particular features of the tasks – for example, the degree of openness; the type of solution sought; and the manner of response expected.

The polarisation of pupils' out-of-school experiences, as evidenced in their chosen hobbies, pastimes and choice of reading and television (see Johnson and Murphy 1986, for a discussion), indicates some of the ways in which perceptions of gender appropriate behaviours are constructed. Such perceptions lead girls and boys to develop the different interests and attitudes to science demonstrated in the surveys. These in turn affect how they engage with school science and their subsequent achievements in it. How pupils interact with science also depends on the image of science that is represented to them in their culture. The uniformity of gender differences across countries gives support to the contention that science has a masculine image in many cultures. The Thai results reinforce this.

Messages about girls' and boys' liking for and achievement in practical work remain unclear. The APU results indicate that girls are well able to handle a range of practical situations and indeed do better than boys in practical tests of observation. The Thai research also demonstrates girls' competence in practical work. The difficulties girls appear to have relate to more traditional curriculum approaches to experimental science. For girls the approach to practical work, the purpose it is seen to serve and the types of problem addressed by it all influence their performance.

The factors which influence gender effects operate throughout pupils' schooling and appear to be particularly critical for their learning in the intermediary school years (ages 13–14). We now take a closer look at the research results to try to establish potential sources of gender differences in science achievement.

Differences in experience

Kelly (1987) found little relationship between previous science-related experience, subject choice and achievement. The APU results paint a different picture. In the APU science surveys boys and girls at ages 11 and 13 perform at the same level on assessments of the use of apparatus and measuring instruments. These tasks were practical and pupils' performance was judged largely by their actions. When performance on individual instruments is considered, girls as a group do significantly less well than boys on certain ones. Boys are better able than girls to use hand lenses and stopclocks at all ages, and microscopes, forcemeters, ammeters and voltmeters at ages 13 and 15. Pupils in the surveys were asked what experience they had, out of school, of the various measuring instruments used. The results show that boys' performance is better than girls' on precisely those instruments of which they claim to have more experience. The different experiences of pupils affect not only the skills they develop but also their understanding of the situations and problems where their skills can be used appropriately. To plan effective classroom strategies it is necessary to focus on both the nature of the differential experiences and their consequences. For example, boys are better able to use ammeters and voltmeters yet they do not have experience of these outside of school. They do, however, play more with electrical toys and gadgets than girls. Such play allows boys to develop a 'feel' for the effects of electricity and how it can be controlled and manipulated. This is an essential prerequisite for understanding how to measure it. For girls to overcome this lack of experience they need to be faced with problems whose solutions, as they perceive them, require certain measurements to be taken. In this way girls will select instruments themselves and engage with them purposefully. This is a very different curriculum strategy to simply encouraging practice with instruments.

Another example of the influence of different pastimes relates to girls' well-documented lack of experience of tinkering and modelling activities. A common strategy employed to overcome this is to provide young children with Lego to play with. Teachers, however, are often discouraged by girls' apparent failure to engage with it. When boys play with Lego they do so in a purposeful way. Their play allows them to establish the link between their purposes and the potential of Lego to match them. It is this relationship that girls need a chance to explore. To facilitate this teachers have to identify the problems that girls find motivating in which Lego can serve a useful function. The same holds true for other areas of girls' inexperience. Erickson and Farkas (1987) investigated gender differences in 17-year-old students' responses to science tasks. They found that females only drew on their school experience whereas males were able to draw on a combination of formal experience linked to their everyday 'common-sense' knowledge. This gave the males an advantage in generating scientific explanations. They also report on females' negative responses to science tasks related to their lack of confidence and fear of handling practical equipment such as bunsen burners and electrical circuits. Girls' aversion to electrical matters is a well-established phenomenon which persists beyond schooling.

When faced with unfamiliar situations it is natural to feel uncertain about them. Girls' lack of certain experiences means that they approach some learning situations in science with diffidence and fear. In the APU surveys one assessment included open-ended practical problem-solving tasks set in a variety of contexts. More girls than boys react negatively to overtly scientific investigations. They express a low opinion of their scientific ability and feel generally unable to respond to such tasks. If the same problem is set in an everyday context and such scientific equipment as measuring cylinders and beakers are replaced with measuring jugs and plastic cups these same girls feel competent to tackle it (DES 1989b).

However, such a strategy has its drawbacks. Faced with an apparently everyday problem both girls and boys, quite reasonably, seek everyday solutions. Consequently they tend not to control variables or to collect quantified data. This suggests that the confidence of girls in practical work might be enhanced by working initially in everyday contexts but with problems that can only be solved using some degree of rigour. Hence there is a need, understood by the pupil, for scientific equipment.

Differences in ways of experiencing

Research into gender differences (Chodorov 1978) has related the different patterns of nurturing that many girls and boys receive to the different values and view of relevance that they develop. These lead them to look at the world in different ways. As a result children come to school with learning styles already developed and with an understanding of what is and is not appropriate for them. What they judge to be appropriate they tackle with confidence, what they consider alien they tend to avoid.

Although girls outperform boys overall on practical observation tasks they do less well than boys on observation tasks which have a 'masculine' content – for instance, classifying a variety of different screws. The same situation was noted in pupils' performance on practical investigations. When offered a choice of investigations, girls who were competent problem-solvers rejected the one with an electrical content as she 'did not understand electricity'. This was in spite of assurances that the solution did not depend on any specific understanding of electricity. Boys, on the other hand, tended to reject the domestic-orientated tasks. They did so not because they regard the tasks as outside their domain of competence but rather outside science. They have a restricted view of the purposes that their knowledge can and should serve.

The results of the APU science surveys show that, irrespective of what criterion is being assessed, questions which involve such content as health, reproduction, nutrition and domestic situations are generally answered by more girls than boys. The girls also tend to achieve higher scores on these questions. In situations with a more overtly 'masculine' content – for example, building sites, racing tracks, or anything with an electrical content – the converse is true. Similar content effects are clear in other survey findings (see, for example, NAEP, SISS and BCSS). Alienation ultimately leads to underachievement as girls and boys fail to engage with certain learning opportunities.

Evidence available from classroom interaction studies indicates that the differences in the nature of the feedback that girls and boys receive about their classwork leads girls to have lower expectations of success and affects the way pupils interpret future experiences (Dweck *et al.* 1978). For example, when girls and boys were presented with an investigation based on content they had already met in class the boys felt confident that they knew the answer, while the girls were inhibited by the belief that they should know the answer.

Girls and boys do appear to experience practical work differently. Randall (1987) looked at pupil–teacher interactions in workshops and laboratories. She found that girls had more contacts with the teacher and of longer duration than boys. However, the girls' contacts included many requests for help and encouragement about what to do next. Randall found that teachers, rather than building up girls' self-confidence, accepted their dependence on them and thus reinforced their feelings of helplessness. The combination of girls' timidity and boys' bravado leads to girls being marginalised in laboratories. Effects of this kind will lead to a real lack of skills in girls and to a substantial problem in future motivation.

Differences in problem perception

An outcome of children's different images of the world and their places in it is that the problems that girls and boys perceive are often very different given the same circumstances. Typically girls tend to value the circumstances that activities are presented in and consider that they give meaning to the task. They do not abstract issues from their context. Conversely, boys as a group do consider the issues in isolation and judge the content and context to be irrelevant. An example of this effect occurred when pupils designed model boats to go round the world and were investigating how much load they would support. Some of the girls were observed collecting watering cans, spoons and hairdryers. The teachers assumed that they had not 'understood' the problem. However, as the girls explained, if you are going around the world you need to consider the boat's stability in monsoons, whirlpools and gales – conditions they attempted to recreate.

In another situation pupils were investigating which material would keep them warmer when stranded on a mountainside. They were expected to compare how well the materials kept cans of hot water warm. Again, girls were seen to be doing things 'off task'. For example, they cut out prototype coats, dipped the materials in water and blew cold air through them. These girls took seriously the human dilemma presented. It therefore mattered how porous the material was to wind, how waterproof and whether indeed it was suitable for the making of a coat.

These examples of 'girls' solutions are often judged as failure either because their problems are not recognised or because they are not valued. Such responses to girls' perceptions of problems means that not only do they typically not receive feedback about their actions but any feedback they may receive will require them to deny the validity of their own experiences. This is a deeply alienating experience.

Another consequence of boys' and girls' different outlooks on the world is that they pay attention to different features of phenomena. A review of the APU item banks reveals that boys more often focus on mechanical and structural details, which reflects their greater involvement with modelling and handling mechanical gadgets both in and outside school. Girls attend to colours, textures, sounds and smells, data boys typically ignore. Consequently, girls always do better than boys, for example, on tasks concerned with chromatography irrespective of the scientific understanding demanded. It is easy to see how these differences go some way to explaining pupils' levels of achievement in physics and chemistry. Girls can deal with mechanical details just as boys are able, for example, to distinguish smells, but their different experiences result in alternative values and perceptions of relevance. When observing or interpreting phenomena pupils will deal with different selections of data unless teachers are alert to these effects.

Earlier mention was made of pupils designing model boats to go round the world. This task, along with one asking for the design of a new vehicle, was given to pupils aged from 8 to 15. The pupils' designs covered a wide range but there were striking differences between those of boys and girls. The majority of the boats designed by primary and lower secondary school boys were powerboats or battleships of some kind or other. The detail the boys included varied but generally there was elaborate weaponry and next to no living facilities. Other features included detailed mechanisms for movement, navigation and waste disposal. The girls' boats were generally cruisers with a total absence of weaponry and a great deal of detail about living quarters and requirements, including food supplies and cleaning materials (notably absent from the boys' designs). Very few of the girls' designs included any mechanistic detail.

The choice of vehicle design and purpose also varied for girls and boys. Many boys chose army-type vehicles, 'secret' agent transport or sportscars. The girls mainly chose family cars for travelling, agricultural machines or children's play vehicles. The detail the pupils focused on was generally the same as before. Where boys and girls chose the same type of vehicle they still differed in their main design function. For example, the girls' pram design dealt with improving efficiency and safety. The boys' pram was computerised to allow infants to be transported without an accompanying adult.

These very different perceptions of the same problem reflect the way girls are encouraged to be concerned with everyday human needs. It was noteworthy that when the pupils had to focus on an aspect of their design the boys had few difficulties abandoning their initial design details. The girls, however, retained their details, which often made it difficult for them to pursue the teachers' focused investigation.

If a problem is perceived to be about human needs then it is not a simple matter to reject that perception and to focus on a more artificial concern, that is, the learning of a specific subject outcome. Yet this is commonly assumed to be unproblematic in science. Furthermore, an ability to do this may only reflect a limited and uncritical view of problems and problem-solving strategies. All pupils will benefit if some attempt is made to relate the focus to the larger problem and to discuss individual perspectives of relevance.

Pupil-friendly practical work?

Rennie (1987), reporting on a study of 13-year-olds in Western Australia, found that the relative inexperience of girls can be overcome in an activity-orientated style of science teaching. Similarly, the science curriculum of Thailand which appears to support girls' learning is described as activity-based, learner-centred, and focused on novel aspects of inquiry and discovery. Girls performed at a higher level than boys across the ages on the APU observation tests. The tasks used in these tests also tend to be novel and allow pupils to test out their own hypotheses before reaching a conclusion. It was found that girls' performance was significantly higher than boys' on such tasks, but particularly so when the tasks were both active and open. Furthermore, the nature of the task can override the content effects discussed earlier. For example, an observation task involving various electrical components in a circuit was popular with boys but not with girls. Boys' performance was also significantly higher than girls. Yet when the performance of girls who liked the type of task, in spite of the content, was compared with boys who similarly liked the task there were no gender differences in performance. These findings lend some support to the view of what a gender-inclusive science curriculum might be like.

It is often concluded from discussions of gender differences that girls need to address problems that reflect their social concerns. Yet the observation tasks described are not untypical science examples. It is important here to distinguish between accessibility and conditions for learning generally. The tasks used in the APU study are accessible because they allow pupils to formulate and test their own hypotheses. It is up to the pupil to accept or reject them on the basis of their own data. It is the pupils' views of relevance that prevail. However, in learning situations, if the new knowledge pupils acquire is to be linked to their existing knowledge then problems do need to be set in the context of human experience and needs. These provide the framework for girls to make sense of the problems posed and the motivation and purpose they need to continue with the complexity of learning in science. Of course, the same should obtain for boys.

Summary

Practical work can play a crucial role in combating gender differences. As the discussion has pointed out, it matters what type of practical work is encouraged. Strategies for changing the aims and nature of science learning to take account of gender differences must be applied thoughtfully. There is little point in attempting to extend the accessibility of science if in the process other groups of pupils become alienated. Classroom strategies will have to take account of boys' and girls' present preferred styles of working and interests as well as providing opportunities for them to reflect critically on them. Many of the tactics that need to be employed to facilitate girls' scientific learning are merely examples of good teaching practice anyway – for example, setting tasks that allow all pupils to express their interests and understandings in a manner that is appropriate for them. However, it follows that tasks of this kind must be fairly broad and general, therefore that pupils will perceive different problems and different solutions within them. Of course, this has always been the classroom reality but the intention now is to make it an explicit and understood aspect of teaching practice. Consequently, there will be a need to adopt a more flexible approach to curriculum planning. A much broader view of the potential learning outcomes in any one lesson or module will have to be accepted by teachers. Similarly, consideration will have to be given to the numerous learning pathways that pupils can follow in acquiring particular scientific understanding.

This is necessary because teachers have to focus pupils' learning and need to be aware of when and how to do this in a way that enables pupils to see the point of the focus and thus continue with their learning. The teacher's role will be even more demanding than at present if such strategies are adopted. On the plus side they will be rewarded by the commitment and achievements of pupils who hitherto have found it difficult to make sense of science. These pupils at the moment not only remain scientifically illiterate but spend a lot of their time in school pursuing purposelessness and meaningless activities.

Note

[1] See also Toh's works in Singapore reported in this book on pp. 89–100.

PART V

Complements to practical science

Simulation and laboratory practical activity

Vincent Lunetta and Avi Hofstein

Laboratory practical activities have long played a central role in the science curriculum. Yet, to conduct practical activities well requires energy, time and resources. One of the ways to improve the science curriculum is to complement school laboratory activities with simulations. But what is a simulation?

The term 'simulation' has been used in several ways in the natural and social sciences. To simulate means to imitate a real system such as an economic system or a process like the flow of fluid through a pipe. Often the simulation or process operates on the basis of a mathematical or logical model. The model is intended to imitate the original faithfully, but generally with less detail (Lunetta and Peters 1985). Interacting with an instructional simulation can enable learners better to understand a real system, process or phenomenon. Both practical activities and instructional simulations enable students to interact with models of reality. Within purposefully contrived settings, both can enable students to confront and resolve problems, to make decisions, and to observe effects.

Simulation has been a part of science education for a very long time. Mechanical orreries were used in the eighteenth century to simulate the relative motion of planets in the solar system. Two- and three-dimensional physical models have been used for many years to represent objects not easy to observe directly such as molecules or human organs. Computer-based simulations are a relatively new resource in science education, and we have only begun to explore their potential. Emerging technologies like video discs, when combined with interactive computing, have great potential to serve as valuable resources in science teaching and learning.

Laboratory practical activities in school science, often referred to as 'experiments', are themselves simulations of scientific practice. 'Experiments', 'labs', or 'practicals' have had a central place in science teaching. There is an expectation that such experiences will enable students better to understand the nature of the scientific enterprise as well as to develop problem-solving skills, to grow in understanding of scientific concepts, and to experience connections between theory and practice. Yet, the designers of school practicals select from arrays of options and generally present to students only a small subset of problems and variables. Even in 'high-inquiry' labs, the teacher limits the scope of the problem. Laboratory activities are generally constrained by school realities such as 40-minute laboratory periods, safety, budgets and resources. The teacher controls lab equipment, space, materials, and even the amount of measurement error that can be tolerated.

Good practicals and good simulations engage students in dynamic problem-solving and inquiry. Laboratory practical activities have the distinct advantage of enabling students to work with real materials and phenomena and to experience their biological and physical environments. Students need direct experiences with organisms, materials and phenomena. Research on learning suggests that experiences with real materials are an essential element in cognitive development. Representations are often misunderstood by learners, and often they are inadequate substitutes for work with material objects. Since simulations provide experiences with representations of reality, it is not appropriate to replace important experiential work in science with simulations. However, it is appropriate to identify time-consuming lab activities that do not provide optimal learning experiences and convert these labs, or elements thereof, into improved simulation activities. To promote meaningful learning in science, the challenge is to develop an optimal mix of 'wet' and 'dry' labs, of laboratory activities and simulations.

In school science, we want students to learn problem-solving and discovery skills in the lab. Expectations in most science courses also include the learning of an array of scientific concepts as well. We generally do not expect students to discover most scientific concepts in the school laboratory. If they are to reach the frontiers of a field or to acquire an enlightened overview, one of the tasks of the science teacher is to help students 'find the shoulders of the giants'. For students, simulations can increase dynamic encounters with concepts and systems. Simulation can be much more than a substitute for conventional laboratory practical activity. Simulations, especially computer simulations, can be designed to provide meaningful experiences that are often not possible with real materials in introductory courses. Simulations can engage students in interactions with problems or models that are too complex, dangerous, expensive, fast, slow, large, small, or too time- or material-consuming.

In the physical sciences, for example, conceptual understanding can be enhanced through experiences with simulations of the Bohr model, nuclear or chemical reactions, the kinetic-molecular model or the nature and flow of electric charge. In the life sciences, simulations can help students examine the growth and development of populations over long time intervals, decode the structure of complex molecules like DNA, trace the steps of complex processes like photosynthesis, and investigate the effects of parental genetic traits on successive generations of offspring. Many important experiments are not performed by biology students because it would take months or even years to collect the necessary data in real time. Investigations in genetics, for example, can take generations before sufficient results are obtained. Computer simulations of genetic phenomena can reduce significantly the time for obtaining information, and hence they can complement practical work in learning about living organisms. Needless to say, the investigator in the simulation also does not have to feed, maintain, control, and observe the living organisms. In terms of the goals for the school lab, this is a mixed blessing. While the simulation may enable students to design experiments and to collect good data over many generations, they do not have the opportunity to experience problems associated with maintenance of the sample. They do not see their organisms die due to experiment-induced or natural causes, they do not experience contamination of the sample, and they do not have to worry about life support for their organisms, 24 hours per day, seven days per week.

In the environmental sciences, complex systems or processes such as food chains, energy consumption, or waste disposal can be simplified and simulated for student investigation. Appropriate environmental simulations can enable students to see connections between school science and the choices and decisions they must make as adults. Informed citizens must be able to use their knowledge of science in resolving significant and complex problems relating to quality of life, disease, hunger, and socio-political concerns. Simulations can assist in helping students move beyond simple scientific models that are isolated from social context to more socially relevant applications of the sciences. Simulations can also enable students to explore microworlds (for example, genes and electric charges) and macroworlds (for example, galaxies and populations) that have not been accessible as a dynamic resource in science education prior to the advent of high-speed computing.

Modes of simulation

To clarify the use and effectiveness of simulation techniques that complement practical work in science education three relevant modes of simulation are outlined in Table 12.1 and discussed here. (For descriptions of additional modes of simulation, see Lunetta and Hofstein 1981.)

Table 12.1 Modes of simulation

Mode	Source	Media (with examples)
1	Secondary sources	photographs, films, videotapes, data tables, graphs
2	Physical models and analogues	three-dimensional models (bonds and atoms; molecular structures; organs; kinetic-molecular vibrations), ripple tank (light waves), colliding coins or ball bearings (nuclear collisions), steam tables, planetariums, aquariums
3	computer-controlled or media-controlled simulations	simulations on computers and/or video discs

Secondary sources

Secondary sources (such as films or pictures) can be used in some school laboratories as sources of data. Students can develop measurement techniques and make observations directly from photographs or from projected images. For example, *Project Physics* (Rutherford *et al*. 1970) includes activities with a series of film loops which show phenomena like colliding freight cars, a pole vaulter in action, and a boat crossing a swiftly flowing river. Students take measurements and collect data directly from the projected images. In introductory labs in astronomy, data are often gathered from photographs of stars or spectra.

Computer or video simulation can be especially appropriate when experiments are difficult, time-consuming, or dangerous to perform or when the equipment required is too massive, microscopic, complex or expensive. In chemistry, simulation can be appropriate when reactants explode, when chemicals are poisonous, or when reactions are too fast or too slow. Ben-Zvi *et al*. (1976a) used filmed experiments to circumvent shortages in laboratory equipment and facilities. In their study, students collected data directly from instrumental readings portrayed in filmed sequences. Research findings have shown that well-designed films of experiments can serve as a viable alternative that enhance some, but not all, student learning outcomes. The advent of interactive video disc technology provides opportunities for the development of new secondary source simulations that may become powerful resources for science education.

Analogues

Some laboratory handbooks provide practical activities that are intended to be analogous to the primary phenomena being studied. In the PSSC investigation 'Simulated Nuclear Collisions' (Haber-Schaim *et al*. 1971), for example, students investigate the collisions of coins in an activity that simulates collisions of nuclear particles.

In teaching about the kinetic-molecular model, apparatus with oscillating balls has been used as a mechanical analogue that can facilitate the study of concepts such as molecular motion, diffusion, pressure and the gas laws. Ripple tanks have been used to study the properties of water waves which serve as analogues of radio waves, light, sound, and other wave-like disturbances. Ripple tanks provide a dynamic medium for simulations that help students study the nature and properties of waves. The water waves can be used as analogues of waves in media that are inaccessible to human senses.

Computer-controlled simulations

Simulations in this category are those in which the information provided to students in response to their queries is controlled by media such as computers and video discs. The electronic computer is a very valuable instrument in the modern science laboratory as well as in the school laboratory. In spite of the diversity and depth of computer applications throughout contemporary science and science education, instructional *simulation* is the only application of computers that will be discussed in this chapter. We shall describe two kinds of computer control for instructional simulations.

1 *Mathematical models*

Mathematical equations stored in computer software can model a scientific system. Programs simulating molecular vibrations enable students to observe the effects of variables such as temperature, pressure and particle mass on animated visuals depicting the movements of molecules in a container. Students can also examine graphics to study effects on molecular velocities and on kinetic energies. Experiences with the simulations and the accompanying tutorials facilitate development of a scientific model of an ideal gas. *Acoustics* (Computers in the Curriculum Project 1983) is another example of a simulation controlled by a mathematical model. This program contains a model of the emission, absorption and reflection of sound in a room with four walls.

Catlab is a computer simulation to support instruction in genetics (Simmons and Lunetta 1987). It is an example of a simulation in the life sciences that is driven by mathematical models. *Catlab* simulates the inheritance of coat colour and coat pattern in the domestic cat. It provides students with opportunities to generate and test hypotheses, to collect data, and to analyse and interpret these data. In *Catlab* students can vary conditions in their experiments. They can control the problem to be investigated, the sequence of their investigation, and the starting and finishing point. In conducting a series of simulated experiments, students will practice many practical and problem-solving skills. They will also develop and act on a framework of structured concepts in genetics. Research data suggest, however, that these investigative skills and concepts are *not* developed by most novices using *Catlab* without sensitive intervention by teachers. Students can become lost in a sea of variables and data in an excellent simulation like *Catlab* as well as in many laboratory activities if they do not receive assistance in controlling variables and in sharpening hypotheses and investigative techniques. In a good learning environment and with sensitive instruction, experiences with *Catlab* can enhance learning and the development of investigative skills. Students also enjoy the experience.

2 *Static data*

Some simulations present or allow access to data collected by others in earlier scientific studies. With this kind of simulation, students can interrogate the data base and search for regularities and order. In the process, they can acquire new information, improve their conceptual understanding, and develop skills in analysing data. One example of a program of this kind is *Birds of Antarctica* (Elizabeth Computer Centre, Tasmania, Australia, reported in Anderson 1984). The data base consists of observations of sea birds made by scientists during a voyage to Antarctica; it also includes time, date, ship's position and meteorological information. A student can interrogate the data to acquire information about the location of certain birds. From the data, the student can construct models and knowledge about habitat, behaviour, and descriptive statistics about the birds. In this particular simulation, students cannot change the conditions of the experiment nor manipulate experimental variables. However, they can gather and analyse data, identify patterns, and formulate and test hypotheses within the constraints of the scientific data available to them.

Each of these three modes of simulation enables student users to learn concepts and to develop scientific inquiry skills. Appropriate lab activities and appropriate instructional simulations can promote learning in science.

Attaining instructional goals with labs and simulations

Science educators have expressed the view that the uniqueness of the laboratory lies principally in providing students with opportunities to engage in scientific investigation and inquiry. The laboratory is 'not just the place for demonstration and confirmation but rather the core of the science learning process' (Shulman and Tamir 1973). Table 12.2 presents an outline of intended learning outcomes for laboratory activity. While conventional practical activities play a central role in promoting these goals, relevant learning outcomes can also be promoted with a variety of related school science experiences. Long-term projects and field trips are among the activities that can complement labs and simulations to develop concepts and skills and provide more holistic understanding of the scientific enterprise.

If we look more narrowly at specific practical activities and simulations within a course of study, it is clear that no single, brief 40-minute experience in the school lab can or should promote all goals simultaneously because of limitations in time and resources. Thus, it is useful to identify the specific subset of goals or objectives that

Table 12.2 Goals of laboratory activity

Domain	Goal
Cognitive	Promote intellectual development
	Enhance the learning of scientific concepts
	Develop problem-solving skills
	Increase understanding of science and scientific method
Practical	Develop skills in performing science investigations
	Develop skills in analysing investigative data
	Develop skills in communicating
	Develop skills in working with others
Affective	Enhance attitudes toward science
	Promote positive perceptions of one's ability to understand and to affect one's environment

Table 12.3 Phases of practical work and modes of simulation

Phase of practical work	Mode of simulation		
	Secondary source	*Analogues*	*Computer-controlled*
Planning and design	−	+	+
Performance	+	+	+
Analysis and interpretation	+	+	+
Application	−	−	+

+ Student engagement is probable.
− Student engagement is unlikely.

is especially appropriate for a particular lab or simulation. Specific labs and simulations can emphasize one or more of the following phases of practical work. In the *planning and design* phase students formulate questions, predict results, formulate hypotheses to be tested, and design experimental procedures. In the *performance* phase, students manipulate materials, make decisions about investigative techniques, observe, and record data. The performance phase is the most commonly emphasised phase in school lab activities. In the *Analysis* phase, students process data in various ways. They may put the data in tubular form, graph it, explain relationships, and search for generalisations. Analysis should also include the development of skills in discussing the accuracy of data and in outlining assumptions and limitations of the experiment. Finally, in the *application* phase, students behaviours go beyond the results of the particular investigation to make predictions in new situations and to formulate new questions to be investigated (Lunetta and Tamir 1979; Kempa and Ward 1975).

Each of the three modes of simulation outlined in the preceeding section enables students to engage in some or all of these phases of practical work. A summary of the simulation modes and the phases of practical activity they best promote is presented in Table 12.3. Normally, with all modes of simulation, data will be gathered, analysed and interpreted. Well-written computer simulations can be sufficiently open to enable users to generate varied hypotheses and to plan and use alternative hypothesis testing strategies. Well-written programs and accompanying video sequences can also enable the student to use the concept(s) in multiple simulated environments and *applications*.

Research on the effectiveness of instructional simulation

Are the goals for laboratory teaching and learning achieved while using simulations? To answer this question properly, there is a need for data from appropriate research. Researchers have investigated the effectiveness of laboratory work in science education, and the findings of these studies were critically reviewed by

Table 12.4 Comparison of learning outcomes, between simulation and practical work

Reference	Simulation	Concept development	Practical abilities and problem-solving skills	Attitude
Kempa and Palmer (1974)	Watching video tapes on manipulative skills in chemistry	Sim > Lab	Lab > Sim	NR
Ben Zvi *et al.* (1976)	Filmed experiments in chemistry	NS	Lab > Sim	Lab > CSL
Bork and Robson (1972)	CSL in college physics	NR	NR	CSL less interesting for low-ability students
Lunetta and Blick (1973)	CSL in high-school physics	CSL > P&P Sim > Lab	NR	CSL > Lab > P&P Sim
Caven and Logowski (1978)	CSL in college chemistry	NS CSL adequate for low-aptitude students	NR	NR
Kinnear (1982)	CSL in college biology (CATLAB)	CSL made it easier to visualise and solve problems	NR	NR
Choi and Gennaro (1987)	Computer simulation of volume displacement junior high school	NS (test on volume displacement)	NR	NR
Bourque and Carlson (1987)	CSL in school chemistry	Mixed results	Lab gave more realistic view of 'trial and error process'	Positive attitude toward CSL

Table 12.4 (*continued*)

Reference	Simulation	Concept development	Practical abilities and problem-solving skills	Attitude
Eisenkraft (1987)	CSL in high-school biology	CSL > Lab CSL helped discovery of functional relationships. Students' abilities interact with treatment.	NR	NR

CSL = Computer simulated labs
NR = Not reported
NS = No significant difference in outcome measures
P&P = Paper and pencil
Note: the symbol '>' should be read as 'more effective than'.

Blosser (1981), Hofstein and Lunetta (1982), and Bates (1978). The reviewers concluded that the issues are complex and that researchers have not comprehensively examined the effects of laboratory instruction on student learning and growth. Thus, inferences must be made cautiously, but there are data and experiences to suggest that the school laboratory does have an important role to play in science education. Similar data and experiences suggest that simulations also have an important role to play in science teaching and learning, but research on the effects of instructional simulation has been even more limited. The lack of excellent simulation software in schools at this time complicates the problem even further.

Bates (1978) is one of several scholars who have called for more systematic comparison of the laboratory with other modes of science instruction. He suggested more careful assessment of what could be accomplished in the school lab that could not be accomplished as well by less expensive and less time-consuming alternatives. A small number of studies have compared the effects of laboratory simulation with more conventional practical work in the lab. Table 12.4 summarises the results of such studies. The studies suggest that instructional simulations were generally not as effective as hands-on experiences in promoting manipulative skills. On the other hand, most studies have found that computer-simulated labs were at least as effective as conventional lab work for promoting concept learning as measured by paper-and-pencil tests. In some of the studies simulations were more effective than conventional labs in promoting concept learning. In general, the simulation activities took considerably less time than did the activities in the school laboratory, and students' attitudes towards laboratory activities and simulations tended to be positive. It is important to point out again that the research studies reported in Table 12.4 assessed relatively narrow information and concept learning outcomes.

In general, they have not assessed some of the important expectations for the school laboratory and for simulation, such as the development of problem-solving skills, attitudes toward science, and understanding of scientific processes (Krajcik, *et al.* 1986). New research is needed to provide a more sound basis for instructional practice.

Implications for teaching and curriculum

Laboratory practical activities and simulations are important resources in school science. Like any other resource, teachers should design and employ them to facilitate important instructional goals. Experiences with even the best labs and the best simulations will result in limited growth for students when teachers are insensitive to goals, to student needs and readiness, to the relevant concepts, and to appropriate teaching strategies. If a teacher's goal is to promote the development of problem-solving skills and not just the learning of facts, this goal should be apparent in what the teacher says and does and in the way the teacher evaluates. Analyses of laboratory handbooks and the limited numbers of simulations now available provide evidence that there are great gaps between the goals espoused for science teaching and the kinds of activity students are generally asked to perform in laboratory and simulation activities. Lunetta and Tamir (1979) wrote that in spite of attempts to reform curricula, students work all too often as technicians in 'cookbook' activities in which they use lower-level skills; they are seldom given opportunities to discuss hypotheses, to propose tests of those hypotheses, and to design and then perform experimental procedures. Contemporary data suggest that the situation has not improved in the intervening decade. The large disparity between goals and practices has been a latent factor in the mixed research findings on the effects of laboratory activity. If the goals outlined in Table 12.2 are important, then specific labs and simulations can be designed and teaching strategies employed to involve students in using skills consistent with those goals.

When to simulate

It is important for students to have contact with real materials in order to make connections between theory and practice and develop understanding of physical and biological reality. It is also important to ensure that students have an appreciation of the steps involved in conducting research. Thus, when an instructional unit can have a relatively efficient and simple laboratory activity which does not involve extensive instrumentation ('black boxes'), it is important to conduct that activity in the school laboratory and not to simulate it. On the other hand, some basic and important concepts are not especially amenable to direct experimental manipulation due to constraints of time, size, danger, or the lack of appropriate resources. Activities involving such concepts are good candidates for simulation. Examples in genetics and kinetic-molecular theory have been described earlier in this chapter, and there are many more, including fast or explosive reactions in chemistry, fission and fusion in modern physics, and use of radio-isotopes, biochemical reagents, and animals in biology. In teaching sophisticated science concepts, when resources are scarce, good computer simulations may be a better investment than specialised apparatus which may be expensive and relatively unused throughout most of the year.

Some commonly used laboratory activities involve excessive effort and time to set up equipment and to gather data. Others involve apparatus that regularly results in large measurement error. While students should come to understand that error is inevitable in scientific measurements, large measurement error often obscures generalisations in a student's data and may inhibit concept development. In such situations, the supplemental use of simulations may be a more appropriate way to involve learners in active encounters with important concepts than experiences in more conventional laboratory activities.

Teaching strategy

It is appropriate at times to employ simulations and teaching techniques that reduce significantly the complexity of reality, enabling students to work with simplified models of the system they are studying. In a complex simulation like *Catlab*, teachers can suggest that beginning students neglect some of the more complex genetic traits and examine only more simple variables like numbers of males and females in a litter or the size of litters, to cite two extreme examples. Students should understand that a simulation is a simplification of reality. Good teaching with simulations engages students in discussing and examining the similarities and the differences between the model and reality. It is often helpful to have students examine the fidelity of the simulation and discuss how they might create a better simulation. With simulations and with lab activities, it is important for teachers to assess the connections students make between the simulation or the lab and reality. If students are unable to make connections between the simulation or the lab and what it purports to represent, the activity should be modified or replaced.

The use of simulations and practical activities provides special opportunities for students to examine investigative techniques. The question of what is an appropriate sample size, for example, is one that merits attention. In conventional school labs, there is often too little time for sufficient numbers of trials or samples or observations, but discussions about appropriate sampling methods and sample sizes are still in order. While activities in school labs must inevitably be limited, activities with simulations need not be limited by constraints of sampling time or numbers of samples. Simulation runs can be conducted and samples collected very quickly. Thus, the results of sample size and sampling methods can be examined more easily with simulations than is normally possible in school labs, even when data from an entire class is pooled and displayed on a spreadsheet or a shared data table. To promote improved learning outcomes from school laboratory experiences, supplementing labs with simulations and developing an appropriate mix of labs and simulations is an important opportunity and responsibility in contemporary science teaching.

Group size

Labs and simulations can be conducted with large or small groups. In the school lab students are generally assigned to small teams of two, three or four for practical work. When used optimally, the small-group environment enables students to share ideas, to question each other, and to assist each other in understanding and in problem-solving. Research has shown that teaching strategies promoting co-operative learning (Slavin 1980; Johnson *et al.* 1985) can enhance cognitive outcomes, attitudes, social skills and work habits. Both labs and simulations provide unique opportunities in science teaching to engage students in co-operative, small-group interaction that can facilitate learning. Simulations, especially computer simulations, can also be used effectively by individual students. On occasion, simulation software is written so that the computer plays the role of 'lab partner'.

Simulations, like laboratory demonstrations, are appropriate to use, at times, in a large-group demonstration mode. They may be particularly appropriate when a teacher wishes to demonstrate how to conduct a special kind of inquiry before releasing students to work with a lab or with a simulation independently outside class. A teacher may also wish to use a simulation to engage an entire class in observing relationships and formulating hypotheses to explain those observations. Ways to test those hypotheses can then be discussed in large or small groups. Such discussion is an excellent precursor to independent student work in the lab or with a simulation (Simmons and Lunetta 1987). Using a simulation to facilitate post-lab discussion can also be helpful on occasion.

Simulation quality

Many instructional simulations fail to fulfil their potential; they mirror or exacerbate problems we associate with poor school labs. In a good simulation, there is normally a high level of student–model interaction. Good simulations engage students in inquiry. They enable students to act on a dynamic problem frequently. The complexity is appropriate for the student's development, feedback is presented in realistic form and is easy to interpret. Good simulation software frequently incorporates dynamic visuals and graphic representations to aid in the holistic visualisation of problems.

Labs and simulations are important resources that have a central role to play in science education. Instructional simulations are simplifications of reality that complement essential lab activities. They provide new opportunities for scientific investigation and inquiry with concepts and systems that are often not possible or appropriate to investigate with real materials in introductory courses. While simulations are not always suitable substitutes for practical activities in the school laboratory, they can enhance concept learning and provide a sense of scientific practice. In the past ten years, new developments in computing and video technologies have opened new opportunities for instructional simulation, and simulations promise to become a powerful and more widely used medium in school science. Unfortunately, there is a shortage of excellent simulation software. The development of that software provides opportunities for creative effort and substantial progress towards improving science education.

CHAPTER 13

Tackling technological tasks

Richard Kimbell

Introduction

In a book largely devoted to the teaching and learning of science, it may be useful to preface this chapter on technology with a brief discussion about the nature of technological tasks. To the casual observer, there will be many similarities between practical science activity and technological activity, even to the point where observation alone will not be sufficient to distinguish between them. The differences are of *purpose* rather than behaviour, for – whatever the observable activity – the point of technology is to intervene in the made world, to modify or improve it in some way in response to identified needs or opportunities (for a fuller debate of purpose, see Assessment of Performance Unit (APU) 1987). Technology is by definition task-centred, involving the active pursuit of a concrete goal.

The emergence of technology

Technology is a relative newcomer on the education scene, but that has not enabled it to escape from all the traditional pedagogical battles – indeed technology in its present form can only really be understood as the outcome of those battles. The crucial dispute has been between those who see learning in technology as a process or activity to be experienced and those whose preoccupation is with the transmission of a particular body of technological knowledge. Oddly, the dispute was not based on any fundamental disagreement of principle about *what* technology is, but rather on *how* it should be taught in schools and often on *who* should be responsible for that teaching.

Technology squeezed into the curriculum at the interface between science and practical or technical studies. It was inevitable, therefore, that whatever the statements of principle, the reality of the classroom would be based largely on these established traditions. In principle, most shades of opinion recognised from the earliest developments that technology is a task-centred process: 'Technology is a process which can be observed fully only from the inside; it is not a readily definable area of knowledge' (Schools Council 1970, p. 4).

A number of strategies were developed, initially by the Schools Council Project Technology, to support the development of technology in the curriculum, the two most potent being 'project work' and 'structured courses': 'The essence of school technological projects is the full and practical involvement in all stages of the technological design process'; and 'teachers sometimes find it helpful to introduce pupils to Technology . . . by running mini-courses in specific fields; topics such as electronics, control, pneumatics etc., have lent themselves very well to this treatment' (School Technology Forum 1974, pp. 6–7).

While it would be simplistic to claim that the technical studies teachers and the scientists had thereby found their respective roles, there is no doubt that the notion of open-ended project work filled many science teachers with dread, and conversely that the scientific and mathematical demands of the structured courses had an identical effect on many teachers of practical subjects.

The resulting dichotomy had profound effects on the emerging pedagogy of technology courses. The structured courses sought to identify and codify new bodies of knowledge (in electronics, pneumatics, mechanisms, materials, and so on) which could effectively be taught like 'practical' science. Once grasped, they could then be exercised in project activity in which students could solve problems *using* electronics, pneumatics, materials and mechanisms. The 'modular' GCE, CSE and now GCSE courses in technology that have proliferated since the mid-1970s from (among other places) the National Centre for School Technology at Trent Polytechnic, present the archetypal version of this view of technology. They typically involve eight or more modules of study that define different bodies of knowledge and from which pupils will choose three to study one each term in their fourth (pre-examination) year. The fifth (examination) year then demands a major project that demonstrates the application of these understandings to some real-world problem. The flavour of these modules is nicely caught in an enthusiastic account of a micro-electronics module which

> has its own piece of hardware, the MENTA, which is a low cost Z80 microprocessor that facilitates the teaching of machine code programming. It interfaces with add-on modules used for teaching control applications. These modules include analogue to digital converter, temperature control, stepper-motor control, model crane control, and fibre optic communications (Page 1983, p. 19)

Despite the agreement in principle that technology is a needs-driven activity, the pedagogy of these modular technology courses shines through clearly – first learn your science, then find some problem to which you can apply your understanding.

Understanding is not something that emerges through the pursuit of the activity, rather it is a prerequisite of the activity.

While this development was gathering pace in the early 1970s, a parallel development was emerging directly from the craft and design tradition. Here, *skills* were the traditional currency of the classroom or workshop but the underlying approach to learning was much the same; first learn your skills, then use them on a project. Again it was the Schools Council, through the Design and Craft Education Project, that (from 1968) first challenged teachers to shift their emphasis from a preoccupation with the simple acquisition of skills by pupils. The project sought to help teachers develop a broader design-based approach that allowed pupils to identify for themselves what they were going to make, design it for themselves, make it and then evaluate it. While science-based technology was struggling with the problem of how pupils were to acquire the 'necessary' body of knowledge in the physical sciences, the craft and design approach had to come to terms with how pupils were ever to design and make things for themselves without a previously established body of skills.

The real triumph of this project lay in the strategy that it developed towards assessment. Having developed a procedurally definitive route (called designing) that involved investigation, brief, specification, ideas, making and testing, the project team developed systems that enabled teachers to assess pupil performance in these *procedures* of designing. This was a key development for it led to the development of examination courses (always the key to acceptability in British education) that focused not so much on the knowledge or skill that a pupil 'holds' in some abstract way, as on their ability to demonstrate their capability with this knowledge and skill, on task, in a real setting (see, for example, the North Western Secondary School Examinations Board CSE course, 'Studies in Design', of 1972). Craft and design teachers were starting to redefine the focus of teaching and learning in ways that enabled – indeed encouraged – autonomous, needs-driven activity.

> Individuals are expected, as they mature, to solve problems on their own and to make decisions wisely on the basis of their own thinking. Further, this independent problem solving is regarded as one indication of the individual's adjustment. It is recognised that unless the individual can do his own problem solving he cannot maintain his integrity as an independent personality (Schools Council 1975, p. 30).

The scene was set for a decade of development on two quite different fronts. Modular courses in *technology* expanded enormously and grew into A-level courses; British school technology developed innovative ways of helping schools to resource the 'high-tech' needs of these courses and national competitions like Young Engineer for Britain helped develop the public image of this fast-developing field. At the same time *design*-based courses proliferated from CSE to O and A level, lending weight to the idea that the procedures were assessable with a good degree of reliability; the individualised nature of pupils work often involved a diversity of school departments, thus encouraging the idea of broadly based design studies and concomitant cross-curricular co-operation; and in this area, too, national design competitions stimulated public awareness. While Technical and Vocational Education Initiative (TVEI) developments (from 1982) were further raising the level of awareness of technology among teachers, employers and politicians, both the 'design' and the 'technology' strands began to confront the ultimate bastion of the education world – the universities – and while some battened down the hatches against such heretical innovation, the increasing tendency has been towards qualified acceptance.

It was against that backdrop that, in April 1988, the Secretary of State for Education appointed a National Curriculum Working Group to define 'design and technology' as a foundation subject in the National Curriculum.

The confusion of 'process'

Superficially, the difference between the two historical strands of design and technology has been the preoccupation with knowledge, on the one hand, and process, on the other. The progressive reconciliation of the two positions – initially through the GCSE National Criteria, the work of the APU Design and Technology project (APU 1987) and through the National Curriculum Design and Technology Working Group report (Department of Education and Science (DES) 1989a) – is an interesting study in its own right, but one for which there is insufficient space here. Rather, I would like to focus on the developing understanding of the nature of the process in design and technology, for only when this is properly understood can we really hope to develop appropriate teaching strategies.

Traditional 'problem-solving' models represented the process as a number of sequential steps (Figure 13.1). Innumerable variants on this basic idea can be found in the literature and one of the most detailed is in Curriculum Matters 9 (DES 1987b, p. 10), in which there are 15 steps from 'the problem' to 'the solution'. The steps include such things as 'write design brief', 'collect data' and 'model ideas'. The difficulty with these linear models, as I have attempted to explain elsewhere

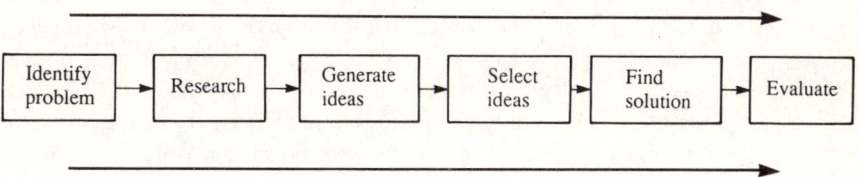

Figure 13.1 Traditional problem-solving model

(SEC 1986, pp. 9–10; APU 1987, pp. 10–11) is that it does not make sense to say that evaluation (for example) only happens at the end of the process, or that ideas are only necessary at a particular point. One has *constantly* to be evaluating, and *constantly* having ideas that demand such evaluation. The models seek to impose order on a messy, confusing and essentially interactive process and the danger is that by imposing order they also impose a degree of rigidity and hoop-jumping that destroys the creative essence of the process.

It is not difficult to see why this came about and the roots of the problem lie in the initial attempts of the Schools Council project to assess performance in the activity.

> In seeking to assess a student's attainment across such a wide range of behaviours . . . [we] . . . need to identify and define the main elements of *behaviour* within a design situation. These elements are presented as educational *objectives* . . . (Schools Council 1975, p. 31; emphasis added)

The behaviours have become objectives, and in the assessment world the next logical step is to ask what *evidence* is acceptable as proof of the existence of these behaviours. The evidence has to be concrete to be acceptable (written down or drawn or modelled) so investigation as an *activity* becomes an investigation *folder* and active design *thinking* becomes a folio of *drawings*. The evaluation *report* at the end of the exercise is the only direct evidence of evaluative *activity* and therefore becomes synonymous with it. The process has become a series of products.

The progressively more detailed attempts to describe design and technology on this model (for example, in DES 1987b), conspire to create a minutely detailed series of products (the brief, the specification) that at best constrain the activity and at worst strangle it.

In confronting this problem, at APU we sought to re-establish the pre-eminence of the activity by recognising that all work in design and technology results from the interplay of thought and action – what is going on in the *minds* of students and what they are doing with tools and materials to confront the *reality* of their emerging ideas (see APU 1987, pp. 12–20). When engaged in a task, ideas (or images) are inevitably hazy if they remain forever in the mind and this inhibits their further development. By dragging them out into the light of day as sketches, notes, models or the spoken word, we not only encourage the originator to sharpen up some of the areas of uncertainty, but also lay them open to public scrutiny in a way that is impossible with internal images. The act of expression pushes ideas forward. By the same token the additional clarity that this throws on the idea enables the originator to think more deeply about it, which further extends the possibilities of the idea. Concrete expression (by whatever means) is therefore not merely something that allows us to see the designers ideas, it is something without which the designer is unable to be clear what the ideas are (See Figure 13.2).

The means of expression often change through this iterative process, for while quick, soft material modelling or sketching is ideal for knocking around hazy ideas, more disciplined and detailed means of expression are required to impose clarity on increasingly complex and detailed thoughts.

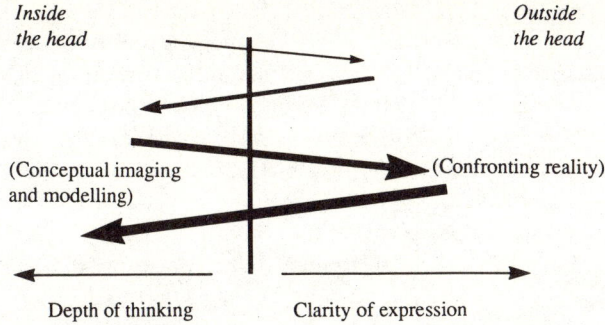

Figure 13.2 An iterative active/reflective description of design activity
NB The increasing thickness of the arrow represents the developing understanding of the task and its resolution.

> we use expressive models to describe and predict how our ideas are developing. A sketch . . . a technical lego construction . . . a sculptor's maquette . . . a stress/strain diagram . . . a computer simulation . . . a prototype . . . an architectural plan . . . All these are 'models' that help us to describe and predict how our ideas will work in reality. The more highly developed our modelling capability is, the more descriptive and predictive our models can be (APU 1987,)

Rather than describing design and technology as a set of preordered behaviours, this model allows us to distinguish between concrete *activities* (taking a photograph, drawing a diagram, constructing a model) which are the expression of thought, and the *intention* that drives it (to investigate something, or evaluate something). These broad guiding intentions change through the life-span of a project as one moves progressively from the identification to the resolution of the task, but often there will be dual or even multiple intentions behind any one piece of activity. As an idea is developed, it must be evaluated, which may lead to a new line of investigation, or even a rethink of the original task. On this model, the whole process is an organic and dynamic entity, not a sterile set of hoops demanding a particular set of behaviours at particular times.

Interestingly, the APU Science team – despite the inherent rigidity of the model they were using to describe scientific endeavour – were drawn to a very dynamic view of 'problem-solving' in science:

> it became apparent very quickly that . . . thinking and doing were interdependent activities. Whilst carrying out the investigations, children were obviously refining their approach to the problem and developing a more complete understanding of it at the same time as developing an appropriate strategy for its solution. This dynamic aspect of the children's problem solving activity seemed to be as important as any other observable outcome (APU 1989a, p. 124).

In the broader context of need-driven tasks in design and technology, the importance of recognising the dynamic interdependence of thought and action is impossible to overstate. For too long curriculum practice and assessment procedure have been dominated by static, behaviour-demanding perceptions of the process. While the simplicity of these models no doubt helped establish them, the price that has been paid is unacceptably high. As an example of one of the greater nonsenses, the 'three ideas' paradigm is worthy of a brief digression.

In craft, design and technology (CDT) examination courses it is commonplace for students to have to demonstrate that they have had a range of ideas in advance of settling on a 'final' idea for development. The concern is that students must demonstrate divergent thinking at the start of a project and (later) convergent thinking as they home in on a solution. In order to demonstrate divergence they must have three different ideas from which they pick one and develop it. Typically, having got an idea as a starting point, students want to get on and explore it to see to what extent it can meet the demands in the task. But the three ideas paradigm demands that they immediately forget the first idea and come up with a completely different idea as another starting point, and then, having got a second, they must find a third. Any similarities between these three ideas renders them 'not different' and accordingly the mark scheme will not credit the student as a fully 'divergent thinker'. Further, having got three different starting points, students are expected (simply on the basis of this first stab at it) to decide which is the most promising idea for development. The students themselves recognise the arrant nonsense of this procedure.

By concentrating on (and specifying) behavioural demands of this kind, many students – and teachers – have become thoroughly disenchanted with what they see as a travesty of real design and technology activity. The tight structuring of behavioural demands has, of course, been justified by the need, first, to make design and technology teachable and, second, to make it reliably assessable. These are both powerful imperatives and at the heart of both we must confront the problem of defining capability. Once we know what counts as being good at it (whatever 'it' is), it is much easier to decide how pupils might be encouraged to develop this capability and how it might be assessed. It is therefore to this crucial question of capability that we must now turn.

Active and reflective capability

At the outset I described design and technology as being concerned with intervention in the made world, a modifying and improving force in response to needs or opportunities. It clearly demands an active, doing, capability. On its own, however, this will not do, for we have all experienced the results of such active capability when it is unburdened by any significant consideration or understanding of the needs that prompt such action. Design failures occur when either of these two elements is missing, when the original need is misunderstood or ignored, or when the emerging product or system has some technical flaw.

Often it is very difficult to distinguish between these classes of failure. Consider, for example, the tin opener that only really works when you squeeze the handles together really hard, but as you do this the metal edge of one of the handles digs painfully into your fingers. What is the source of the failure here? Is it technically bad, or is it that, because the designer is trying to produce a product at a low price (because of an identified market need), some compromise in comfort has to be accepted? Optimising is the name of the game and invariably one is trading *technical* profit or loss against *user* profit or loss. There is never a right answer, just better or worse optimisations and what counts as better or worse depends entirely on your value position (as a retailer, parent, ecologist, manufacturer, or whatever).

These are the two crucial aspects of capability in design and technology. If one cannot fully understand and empathise with the needs of users, then it is very difficult to design for them. It is for precisely this reason that architects are urged to spend a week of their training in a wheelchair. At the end of the week they see the world in a different way. But empathy alone gets you nowhere, except perhaps in terms of a richer social conscience. In design and technology it needs to be allied to the sort of creative, developmental capability that can conceive of new arrangements of materials and functions and new visual, structural, mechanical, or electrical possibilities.

At APU we developed an assessment framework to reflect these twin imperatives and which allowed us to develop a concept of holistic capability. The framework originated in the model of the activity I outlined earlier and it has three ingredients – *conceptual* understanding (inside the head), explanatory or *communicative* facility (outside the head), both of which are held together and given meaning by *procedural* capability in the task (Figure 13.3).

Procedural capability is at the heart of the matter, for in design and technology it is the driving purpose of the *task* that prevents conceptual understanding being merely intellectual detritus, and communicative facility being merely organ

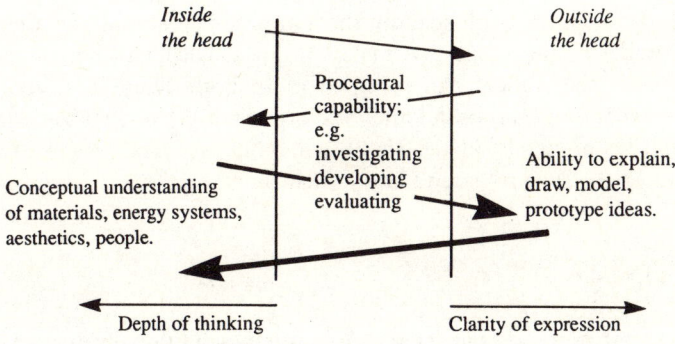

Figure 13.3. Concepts, procedures and manifestations: the axes of the assessment framework

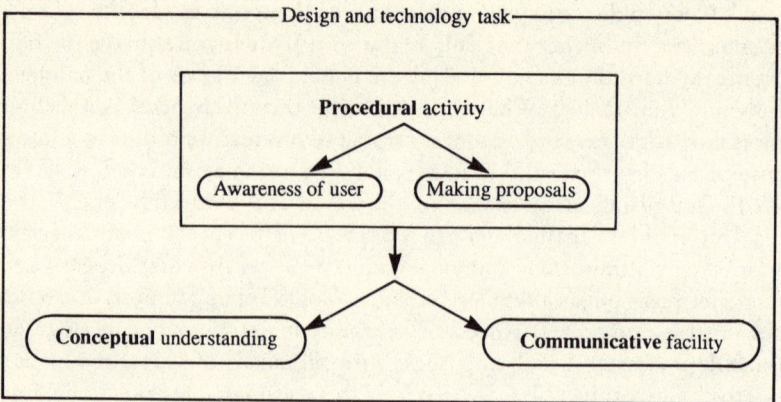

Figure 13.4 Two levels within the assessment framework

grinding (suitable for house-trained monkeys). There is a world of difference between, on the one hand, holding (in some abstract sense) a concept of electrical power or a skill in perspective drawing and, on the other hand, being able to *use* these understandings and skills to some purpose. In design and technology, capability is defined in this latter, active sense. Totally incapable people (in design and technology terms) may hold a rich variety of conceptual understanding or communicative facility. It counts as nothing if the two cannot be welded together and used purposefully in a task. Moreover, it is this active pursuit of a task that enables students to see what more they need to know and find out about, thereby expanding their conceptual platform.

There are, therefore, two levels at which we can see the active/reflective analysis of capability; within the activity as a whole and within the procedural component of it (Figure 13.4). The two levels are, of course, closely related, for an awareness of users depends on understanding about the nature of people and how they live and interact. Similarly, the ability to make proposals depends on the facility of expression that the student can bring to it. In both cases, however, it is the procedural capability (purposeful and task-orientated) that takes the understanding and facility onto a new plane, not only making use – and sense – of both, but also providing the spur to extend and enrich them.

Assessing capability

Assessment strategies in design and technology have traditionally divided the active and reflective modes of working and thereby destroyed their validity as tests of capability. Conceptual understanding, for example, has typically been tested in passive, written examinations and communicative facility through active (but closed-ended) drawing or practical tests of various sorts. For the APU national survey, the logic of our argument forced us to reject this division and devise new approaches to testing. Our view of capability demands that we should not be interested in what students *know*, or what *skills* they possess, but in the extent to which both of these can be *used purposefully* in a task. All our tests are therefore task-centred and devised in ways that enable students (in a very short time) to get inside the demands of the task and display their capability in using and extending their understandings and skills.

The strategies we used to accomplish this were many and various (for example, the use of video material to help students access the context of the task) and there is not space here to discuss them. I have outlined our methods elsewhere (Kimbell 1988, p. 107). Suffice it to say here that the survey, in November 1988, involved the use of 27 different tests on more than 10,000 pupils in approx 700 schools in England, Wales and Northern Ireland, and it has resulted in the accumulation of the world's largest data base on capability in design and technology. We have test performance data of various sorts as well as pupil and school background data that in combination allow us to diagnose the crucial constituents of capability and the circumstances (human and curricular) that seem to be associated with it. We are only now, at the time of writing, beginning to derive findings from the data base, but during the four years of research, including materials trials and a pilot survey in 1987, we have established a number of hypotheses that we believe we shall be able to substantiate. The following sections summarise some of the major issues we have identified in relation to the tasks themselves and the characteristic responses of students.

The nature of tasks

One of the areas of our work that has most taxed us is the problem of creating tasks at equivalent levels of complexity. What is it that makes tasks easy or difficult? Clearly, it is quite possible for a task to involve conceptual understandings (for example, about electrical energy systems) that students do not – at the inception of the task – hold, and equally it is possible for tasks to demand skills (for example, of prototyping) that students do not immediately possess. To that extent, therefore, tasks can appear more or less difficult, but the essentially procedural nature of design and technology enables students to overcome these difficulties. It is commonplace in design and technology tasks for students to seek out (on a need-to-know basis) new areas of conceptual understanding or practical facility according to the demands of the task and any specific lack in these areas cannot in itself account for the difficulty of a task. The question we must ask is what makes tasks procedurally simple or complex.

As I pointed out earlier, the essence of design and technology lies in the optimisation of a solution – creating the best possible balance between all the conflicting priorities in the task. It follows that the more priorities one identifies, the more difficult it will be to balance them. It may not sound difficult to design a new toy for five-year-olds, but if the toy must not only entertain and amuse the child but also be safe, be cleanable, be cheap, be soft, be durable, float, be light, and so on, then one soon has to start trading one quality against another. The more *issues* one can see as being important in resolving the task, the more complex it is to resolve it.

> Evidence from our testing suggests that the very same task . . . is immediately reworked by pupils into tasks of dramatically varied complexity. By deliberately closing their minds to some issues (constraints), pupils can make tasks simpler and equally by recognising a broader range of issues, a simple task can become very complex . . . Often pupils seem unaware that this is happening – and commonly this re-working leaves girls with more complex tasks than boys (APU 1989b, p. 11).

The complexity of a task is tightly bound up with the *context* in which it resides. The context carries with it a pattern of understandings and experiences that students use to get a better grip on the parameters of the task. Decontextualised tasks (like designing 'a can-opener'), have very little meaning because it is the context – or set of circumstances within which the product must operate – that provides the meaning. Designing a can-opener in the context of elderly (and potentially infirm) people in small domestic kitchens is a task that is more immediately understood. But it is a quite different task to designing a can-opener for the kitchens of a huge hotel. The context provides access to the issues that will bear on the task, and these in turn affect its complexity.

Structuring the response to tasks

Because of the complexity of tackling tasks in design and technology, it typically takes a long time from the inception of a project to its completion. Project work for GCSE, for example, can normally be expected to last for anything between a term and a year. While we used a large number of such projects as case studies of performance, the vast majority of our data are derived from shorter structured activities that were targeted at identifying particular combinations of qualities. These structured activities were initially designed to facilitate the collection of as much data as possible in the shortest possible time, but it soon became clear that the different test structures that we were developing were (by virtue of their structure alone) generating different types of response. A typical structure would involve several stages of activity; for example, in an evaluation test these might be:

1 Watch context-setting video.
2 Receive two similar products that typically exist in the context.
3 Identify criteria that would be appropriate for judging them.
4 Identify strengths and weaknesses in the two products.
5 Redesign one of them to overcome noted weaknesses.

6 Design a testing procedure that could give reliable performance data in relation to an identified quality.

7 Explain how measurements are taken and reliability is ensured.

The whole activity would last no more than 90 minutes and would be supported by the video, a teacher's administration guide and a response booklet that guided students through the stages of activity.

Despite the efforts we made to ensure that the activity had real meaning for pupils (for example, through the use of video to set the scene), our initial fear was that the artificiality of such an activity would seriously depress the level of responses. We were therefore amazed when the quality and depth of responses exceeded that which was associated with evaluation activity in real design and technology projects. As one teacher commented, 'that boy has done more thinking about evaluation in 90 minutes than he has done on his major project'. When it is handled thoughtfully, structuring the activity *supports* performance.

But if structure supports (that is, affects) performance, then different structures can be expected to result in different performances. This will, I believe, emerge clearly from the data for in some areas the differences are marked. Consider, for example, the difference between these two structures for the evaluation test;

(a) 1 Video contextualising.
   2 Identification of criteria of judgement for a product.
   3 Presentation of product.
   4 Strengths and weaknesses of product.
   5 Redesign to improve product.

(b) 1 Video contextualising.
   2 Presentation of product.
   3 Design development of a new product to be better than it.
   4 Identification of criteria of judgement for original product.
   5 Strengths and weaknesses.

The only difference is in the position of the *active* designing (stage 5 or stage 3), as opposed to the *reflective* identification and consideration of criteria, but this has enormous consequences for performance, especially in terms of the richness of the criteria identified as important. My earlier analysis of the interdependence of thinking and doing enables us to understand clearly why it happens. This, along with many other structural features of tests, must be recognised as impinging, in a major way, on what we might be tempted to assume is the 'real' performance levels of students.

Influences on performance

I have attempted to indicate in this section some of the difficulties of creating 'standard' tasks and of assuming that performance is a product of student capability alone. In facing these difficulties, we optimised as best we could and created a broad spectrum of test activities that we believed would allow us to report effectively on student capability. We used a range of contexts, tasks and response structures and stratified the survey design so as to control the placement of different categories of student (especially with regard to gender and curriculum experience) within the design. We are now starting to see the patterns of characteristic responses that emerge from the different groups. But the patterns are confusing.

Certainly we can identify gender effects and curriculum effects and ability effects, but in attempting to understand the nature of these effects we are studying the difficult areas where (for example) the characteristic gender effect clashes with characteristic curriculum effect. Girls who do study CDT as against those who do not is a case in point. Even this can be confusing, however, unless we also recognise ability levels and a number of other variables as part of the equation.

We are engaged in a multi-level modelling exercise attempting to describe the nature of capability and – within the limits of our data – explain why it is that particular groups perform in particular ways.

In conclusion

Design and technology is standing at the threshold of major developments that will, in the next few years, place it firmly within the curriculum experience of all pupils aged 5–16. Seldom can there have been such a rapid rise to prominence, for at the outset of the 1980s it did not exist in anything like its present form and in 1970 it was an illusive dream for a handful of pioneers.

The debates and battles of the last few years have resulted in a uniquely process-centred rationale but, while this would be broadly accepted by most practitioners, underlying that general consensus are many serious confusions about the nature of the process and what counts as capability within it. It is these matters that I have sought to address in this chapter for they are the issues that we have had to face in progressively greater detail as we have sought to create new assessment instruments. The closer we got to our final prototypes, the sharper the focus became on the issues that influence their effectiveness and, as the focus sharpened, we were ever more able to see the failings of our materials. But then, as I have attempted to explain, that is the nature of design and technology.

PART VI

The assessment and evaluation of practical science

CHAPTER 14

Principles of practical assessment

Bob Fairbrother

'If I am not teaching, you are not learning'?
'If I am teaching, I cannot be assessing'?

Introduction

This chapter is about strategies, not tactics. It is about principles, not detail. It discusses what assessment is about in general terms rather than what to do in detail. (For fuller details about the assessment of practical work, see Fairbrother 1988.)

Many, perhaps most, teachers have an outlook which embraces one or both of the quotations set out at the beginning of this chapter. I believe neither of them is true. This chapter is influenced by this belief and tries to deal with some of the issues which give rise to them or are a consequence of them. The issues apply to all the teaching and to all the assessing that teachers do, but in the context of this chapter they are dealt with in the assessment of practical work.

In England and Wales teacher assessment in science for formal purposes is tending to become just the assessment of practical work. This is because the GCSE examining groups have combined the two requirements of coursework assessment and practical work assessment into one operation. The National Curriculum being introduced is likely to change this since it requires the teacher assessment of much more than just practical work.

More importantly, perhaps, in the more informal, day-to-day teaching activities, assessment has always involved everything the students do. So in one sense the National Curriculum is not asking for any more than what we have always been doing. In another sense, the fact that the assessment now 'counts' and must be recorded, there are big differences.

Holistic and atomistic approaches

We need to know what education is for before we start teaching, and we need to know what we are teaching before we can start assessing. We ought to be clear about what we mean by 'practical work', how we think it should be incorporated into our teaching, and how it relates to theory. (See Woolnough and Allsop 1985, for a good discussion of the role of practical work in science. See Solomon 1980, for an attempt to look at things from the point of view of children.) Our view of these things influences the kinds of activity that we arrange for our students and the way we assess them.

There is a spectrum of practical work which goes from the general to the specific. At the general end we are concerned with open-ended problem-solving tasks which can last any period of time from a single period to several weeks. Assessment is difficult because it is a non-linear situation. We do not know the 'track' which the students will take because the nature of the problem, the methods of solution and the acceptable solutions are not spelt out. At the specific end we are concerned with a single skill which is well defined and for which the criteria for success are clear. Assessment in this case is easier.

As an example at the general end, let us take the following problem:

Explore a range of floor coverings to see which one is the best for use in a kitchen.

This problem is capable of being tackled from a wide variety of standpoints, and can involve a wide range of expertise which involves other subject disciplines as well as science. Is it possible to draw up a mark scheme? What are the criteria by which one judges an answer? How would a report, written by a consultant who was given this problem, be judged (assessed)?

Assessment becomes easier as the problem becomes more specific because there is much less freedom of choice. You construct the problem to obtain a more clearly defined response; you are able to anticipate possible answers and so draw up a clear mark scheme. For example, the following problem is much more directed; it leaves little room for decision-making although we do not know what the answer is:

This egg timer is designed to time 4 minutes. Check it several times with a stop watch to see whether the sand does take 4 minutes to run through the timer. Give a clear account of what you did and what you found out.

What would a mark scheme look like this time? Perhaps it would include a list of readings (at least three), a description, and a conclusion which relates to the problem. We will not follow this through to the end and deal with the number of marks, specimen answers, and so on. But it raises questions about why we chose those items for a mark scheme, and whether problems like this are suitable for students studying science in school.

At the specific end of the spectrum the task is so clearly defined that it loses its problem-solving nature in all its aspects, except that it may not be easy for the students to do it. Take a task like the following:

Use the metre rule supplied to measure the length of line AB accurate to the nearest millimetre

A   B

The mark scheme can now be simply dichotomous – right/wrong or pass/fail or 1/0 – depending upon whether the answer given is the correct length or not. Again, the question should be asked whether such a task is appropriate for science students and, if so, whether it is a suitable task for an end-point examination (that is, for summative purposes) and whether it is a suitable task for checking teaching (that is, for formative purposes).

Education and examination

There is a strong movement in the direction of open-ended problem-solving in science education. It is relatively new and in England and Wales it is almost impossible to avoid since it is now a part of the GCSE and the National Curriculum. It is seen to be not only a reflection of the way 'real scientists' work but also a good way of teaching science skills and knowledge. This begins to answer the question about why problems like the ones suggested above might be appropriate for students. The choice of task and the items included in a 'mark scheme' reflect a mixture of one's own view of what science should be like, the purpose of the task – whether it is seen mainly as formative (a part of teaching), or mainly as summative (a part of examining) – and the constraints of the system. The 'mark scheme' can have great variety ranging from informal day-to-day feedback to the students to formal examination-type criteria. The 'system' includes the way each institution works as well as the curriculum, the syllabus and the examination.

At the level of teaching students in the laboratory we are concerned with advancing their knowledge and skills. Within the context of a coherent and complete investigation it is possible to concentrate on specific skills such as using a metre rule, lighting a bunsen burner, using a hand lens, recording data, drawing graphs, forming hypotheses, and so on. Many people argue that this is the proper way to deal with specific skills – integrating them in a context, not disintegrating them into separate activities. The egg-timer problem above might be used in order to develop the skill of timing with a stop watch, but there are other skills involved such as tabulating results, communicating and drawing conclusions. If the students are good at all the other skills, then this might be seen as a single-skill task. The metre-rule task is much more clearly single-skilled.

Sometimes we narrow down the demands we make on students so that we can introduce or develop a particular skill or get information about performance in that skill as reliably as possible. A task which concentrates on a single specific skill, such as the line-measuring task above, is likely not to be seen as an investigation, although it might require the students to make some decisions for themselves. (For example, do I put the beginning of the line at the beginning of the ruler or somewhere else?) Concentration on a single specific skill could mean that the skill is seen in isolation. If it is treated out of context and unconnected with any other skill and, in the extreme, unconnected with any reason for developing the skill except for the purpose of obtaining a mark in an examination, then we must question the validity of what we are doing. Many examinations are like this, however; one has only to look at the disjointed jumping from one item to the next in a multiple-choice paper or in some of the old CSE practical circuses to see examples of this practice. When teaching, however, the development of a specific skill might be acceptable because it can be seen in the context of a series of lessons in which, unlike in an examination paper, the purpose of developing the skill can be made clear. Its assessment, for formative purposes, is not seen as a mark-grabbing exercise. This is discussed further in the next section.

Transfer of learning: assessing in context

The choice of task from the spectrum of the general to the specific, and the way a task influences the design of the curriculum, methods of classroom teaching and hence the assessment of student performance are all a function of two things. One is the external, imposed constraint of the examining system which has been mentioned above and will be discussed further in the section on validity and reliability below. The other is more subtle and concerns the extent to which we believe transference of skills takes place. We teach and assess practical skills not as an end in itself but because we believe these skills can be applied in other situations.

Another way of looking at this is that we should be assessing in a way which enables the students to see the purpose of what we are asking them to do. That is, we should put the assessment into some kind of meaningful context. As discussed above, this involves investigations or explorations or open-ended problem-solving which, for the purpose of this chapter, I want to treat as synonymous.

If large-scale transfer takes place, then we can teach and assess generalised practical skills which can be transferred to a wide range of specific problems. If limited transfer takes place, then we have to teach and assess practical skills in more specific contexts. Generalisability and transferability involve the identification and integration in some way of the skills which are required to solve problems. They are thus concerned with both the atomistic and holistic approaches to teaching and assessing practical work which were discussed above.

The world context

An extreme view is that generalised problem-solving is not subject-specific. It is process-based and content-free. Authors such as de Bono (1970), Jackson (1983) and Instone (1988) try to identify stages in the problem-solving process which apply to all problems. Instone suggests:

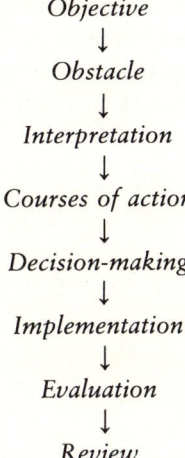

Objective
↓
Obstacle
↓
Interpretation
↓
Courses of action
↓
Decision-making
↓
Implementation
↓
Evaluation
↓
Review

Adopting this approach one is supposed to be able to teach people to solve a wide variety of problems.

The science context

If there is such a thing as process-based content-free problem-solving, then a single problem-solving task might satisfy the requirements of several subjects in, say, the GCSE. Even if the task is not content-free, there may be sufficient commonality for the task to be acceptable in different subjects. In the National Curriculum, however, the tasks have to be set into a science context. Attainment Target 1 (AT1) for Science says:

> Pupils should develop the intellectual and practical skills that allow them to explore the world of science and develop a fuller understanding of scientific phenomena and the procedures of scientific exploration and investigation. This work should take place in the context of activities that require a progressively more systematic and quantified approach, which draws upon an increasing knowledge and understanding of science. (D.E.S., 1989)

So, even if the procedures apply to more than one subject, they must, nevertheless, be put into a science context. Furthermore, the spirit of AT1 is that students should be involved in complete, coherent, holistic activities. The assessment of individual skills can take place but in the context of such activities, not as isolated skills. The emphasis should be on the assessment of the skills while they are being used for some real purpose.

Assessment requires that there should be some model of what we think the students should be doing. The stages identified by Instone mentioned above are steps taken in a linear model of solving a problem. There are several other attempts to describe the process of solving problems.

The Graded Assessments in Science Project (GASP) describes three stages in doing an exploration: *planning, implementing* and *concluding*. These are defined in more detail by 'criteria' which are involved in each stage. A pupil version is available, supported by a Pupil Record Sheet which suggests the pupils give information under the following headings, in which the three stages can be identified:

Title
What I am going to do?
What do I need to use?
How I am tackling the exploration?
What did I find out?
Were there any problems?
Could anything have been improved?

The Science team of the Assessment of Performance Unit (APU) (see, for example, Gott and Murphy 1987) suggests an iterative approach, as shown in Figure 14.1. The iterative part is concerned with changing one's mind in the light of fresh evidence and going back to an earlier stage to reformulate the problem, change the plan, do something different, and so on. In effect all the activities cover much the same ground as in the three stages of the GASP scheme. Sargent (1989) does a detailed analysis of explorations in the Sussex Science Horizons scheme and sees a close parallel with the APU stages. Gott *et al.* (1988) identify two sorts of understanding being required to do investigations – conceptual understanding (concerned with facts, concepts, laws and principles), and procedural understanding (concerned with how science is used to solve problems). Their description of what is involved in procedural understanding is derived from the APU scheme. It seems that the APU has strongly influenced the development of thinking about investigations in the past ten years, and there is a similarity in all the schemes which are proposed for solving open-ended problems.

The basic pattern is that one has to *start* somewhere, *end* somewhere and do something *in between*. Any differences which occur between the different schemes involve the specification of the detail, the order in which the detail comes, and the interrelationship between the detail. For example, a problem might start (emerge) from observations made when trying to solve another problem; it might arise out of spontaneous interest in a phenomenon; it might simply be given to pupils as a part of a closely planned curriculum. Each of these will put different detail into the starting phase. The APU scheme is concerned with an 'imposed' task which is not an integral part of a taught curriculum. The starting point is to find out from the pupils what they think the problem is about, and so is concerned with the pupil's perception of the problem and its reformulation into a form which can be investigated. This is also the emphasis in the GASP scheme which tends to be used as a part of a closely planned curriculum. The start is basically concerned with

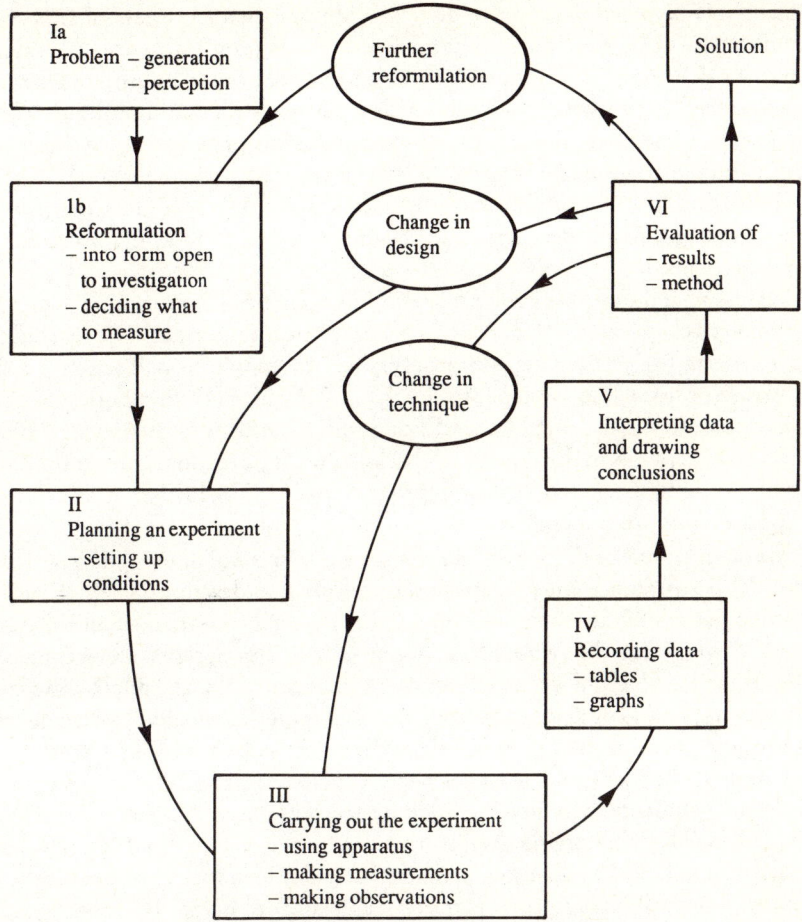

Figure 14.1 An iterative approach to science as a problem-solving activity (after APU)

making decisions about the 'nuts and bolts' of solving the problem such as saying what things will be needed, identifying the variables and making a testable hypothesis. The GASP scheme suggests the possibility of changing the initial plan during the Implementation stage; the APU leaves this possibility until evaluation takes place.

The purpose of these different descriptions is to find a way which validly describes the processes which are used in solving a problem so that pupils can be helped to become better problem-solvers. This is crucial to the assessment process. It may not be achievable in any unique form, not only because problems arise in different ways but also because different pupils will almost certainly make decisions about solving the same problem in different ways. It is unlikely that a description of problem-solving which is valid for 16-year-old pupils with 11 years of experience behind them will be the same as that for five-year-old pupils who are just starting. Even if one does arrive at a unique description, it will be quite complicated in trying to spell out the factors which are involved and putting them into some kind of order which at the same time allows for flexibility and iteration. One is then faced with the difficulty of conveying this to the pupils so that they understand what is involved. Modified descriptions may be needed for different audiences. One can imagine having a simple description for beginners and a more complex one for the more experienced.

As assessment moves more towards criterion-referenced approaches the involvement of pupils in the assessment process becomes more important. The reasoning is that if criteria for attainment and progression are known to teachers, then they should also be made known to pupils. Lock and Wheatley (1989) report on work in connection with the Oxfordshire Certificate of Educational Achievement (OCEA) which involves pupils in their own self-assessment. The National Curriculum, with its strong emphasis on criteria and progression, makes further moves in this direction even more imperative.

If the descriptions of the way problems are solved are valid, then the assessment of pupils' performance must map closely onto these descriptions. If there is a mismatch between what the pupils do and the descriptions, then either the pupils are not doing what they should be doing or the descriptions are wrong. For example, the GASP Pupil Record Sheet sometimes constrains the pupils to work in ways which they find difficult to follow. Sometimes they are not able to decide what they are going to do until they have 'played around' with the problem beforehand. Their desired starting point is to get some experience first.

Assessing students will involve looking at how they are getting on in an individual task and then trying to relate that to some kind of action. We will want to find out strengths and weaknesses so that we can use them to help students take the next step. This will involve looking more deeply and atomistically at what goes on in an individual task. We will not be making broad, dichotomous decisions of the kind: 'Pass; go on to the next task on the ladder. Fail; do another task at the same level.' I think we will be involving students in a holistic process of problem-solving, but doing an atomistic assessment. Occasionally we may have to have some means of bringing together the results of the atomistic assessment in order to report on the progress being made. (In the National Curriculum in England and Wales this has to take place at ages 7, 11, 14 and 16, and also when reporting on individual pupils to their parents every year.) This is discussed further in the section below on validity and reliability.

Progression

Progression in a subject is strongly linked to the ability to transfer one's learning to other situations. Teachers in England and Wales are being encouraged to see the

National Curriculum as a means of enabling students to progress through a subject, and to see assessment as a means of identifying the progress made and of helping further progress to take place. For the first time we have, in the National Curriculum, an attempt to provide a formal link between teaching and assessment. The different levels within each Attainment Target are intended to mark out stepping stones in the progression. It is certain that some of the levels will be found to be in the wrong order or, at least, that some of the statements of attainment which define a level are in the wrong place. I think there never will be a universally correct order because different students will progress in different ways, but we will eventually be able to improve upon the first, rushed attempt.

AT1 is the one most concerned with practical work although the others should be taught through a good mixture of theoretical and practical activities. The idea of progression which seems to be implicit in AT1 has several dimensions, some of which are described in the *Non-statutory Guidance* (DES 1989c). One dimension which is identifiable says something about what a pupil does, for instance:

Ask questions→ Ask questions, hypothesise, suggest and test answers

Other dimensions are concerned with the nature of the tasks, such as:

Concrete situations	→	Abstract situations (hypothesising, using scientific knowledge and understanding, second-hand data)
Familiar/everyday	→	Unfamiliar
Qualitative/descriptive	→	Quantitative (in both collection and analysis of data)
Simple	→	Complex (in scope, design, execution and reporting

This assumes that transference of skills within a subject will take place to some degree otherwise it would be necessary to reteach old skills every time the task changed to one which was more complex or more quantitative. If we have a programme of problem-solving within science, we assume that students will develop a general ability to solve problems in science. (If we cannot do this, then AT1 is a nonsense.) We ask the students to tackle a limited number of problems and hope that they will thereby be able to tackle other problems. It seems reasonable to assume that if students have a lot of experience with problems of a particular kind covering a certain range of concepts and skills, then they will be able to tackle other problems of the same kind and with the same concepts and skills. This is transference *across* the subject involving problems which are similar in their concreteness, familiarity, qualitativeness and simplicity.

But what about transference *up* the subject which enables progression to be made? What is the size of the jump into the 'unknown' that we can expect pupils to make? Is it reasonable to expect pupils to go from concrete problems covering a certain range of concepts to abstract problems covering the same concepts? And then to abstract problems covering different concepts?

The major problem seems to be that of ensuring progression along several dimensions at once. Can two students A and B progress at different rates along the different dimensions? For example, could we have the following at any one time?

Concrete situations..........A..........B.......... Abstract situation

Familiar/everyday ..........B..........A..................... Unfamiliar

Qualitative/descriptive.....A..........B..................... Quantitative

Simple.....................B..........A..................... Complex

That is to say, could student B be ahead of A in coping with an abstract problem, but be behind A in coping with an unfamiliar problem? I think it could happen. In general we will find one student at a lower level than another student in one Attainment Target but at a higher level in another. Looked at in more detail, we will find different students achieving different statements of attainment within the same level. This means that a single task must allow for a spread of attainment in each of these factors; it cannot be targeted at a very narrow range of statements of attainment or levels. We must be looking for differentiation by outcome as well as by alternative tasks.

Validity and reliability

We are frequently caught between the conflicting requirements of validity and reliability. Left to ourselves we would, as good teachers, encourage students to develop their knowledge and skills so that they can apply them to a wide range of situations. Assessment in this context is for formative purposes, that is, identifying where students are and helping them to progress. An assessment which does this is valid for educational purposes. For examination purposes we tend to shift our emphasis. We have to report the results of the assessment to other people, and we are concerned that we are being fair to our students. We want to assess each student in the same context and in the same way so that we can give the same 'mark' for the same behaviour from different students. This concern with fairness leads to an emphasis on reliability and we tend to adopt the methods of the external examining boards.

Practical examinations have been a part of external examinations in England and Wales for many years, particularly at Advanced level. To begin with, they concentrated on those activities which produced a permanent result which could be sent away and marked by an external examiner. Standardised tasks came into being which involved the students following set instructions: assemble the apparatus in this way; do this to the apparatus; take these readings; plot a graph of this against that; find the gradient; from your value of the gradient find such-and-such a constant. Being able to do this kind of examining was a great step forward at the time. It depended on there being a complex infrastructure involving communication between schools, examination boards and examiners; on the availability of apparatus, laboratories and other facilities in schools; and on the existence of trust between all concerned. These did not arrive overnight. However, the results had to be such that they could be written down and marked to a common standard by other people in different geographical areas. This meant that only a limited number of objectives could be assessed reliably in this way. An unfortunate backwash effect was that teaching tended to concentrate on these particular objectives with the result that only that which could be assessed in this way became important. (This was accompanied by theory papers which tended to require the description of standard experiments which had already been done in the laboratory and were described in books of practical experiments.)

Eventually the range of objectives was broadened to include aspects of safety and some independent decision-making on the part of the students. Good tasks which gave the information required were difficult to fit into a 90-minute examination slot. There was also the realisation that a single assessment occasion was not very fair to the students and also not very reliable. The solution was to involve the teachers in the assessment of their own students. Hence teacher assessment and the associated moderation of standards came into being.

The Nuffield A-level Physics Project was one of the earliest to introduce the teacher assessment of investigations. The nature of an investigation is such that one cannot know exactly what a pupil will do. Devising a mark scheme which can be used by a number of teachers in different geographical areas thus becomes rather difficult. The method adopted by the Nuffield team was to suggest 'criteria' related to the following:

1 Applying knowledge and understanding of physics of an appropriate standard at all stages of the Investigation.
2 Showing sensible scientific behaviour.
3 Devising and carrying out simple, effective experiments.
4 Incorporating a variety of worthwhile ideas – even if they are not always successful.
5 Appreciating the meaning of observations and being able to evaluate them critically.   (Harris, 1985)

The examination board issues further information which expands on these criteria and explains how they should be used by the teacher to judge the quality of a candidate's work in awarding an overall grade to the final project, taken as a whole, bearing in mind the given criteria.

It was natural to try to link all this to the normal formative assessments which teachers were doing anyway, and this is where we are now in the current GCSE. The conflict between the formative purposes and the summative purposes of assessment is now quite strong. Many teachers are not able to combine them. Formative assessment involves helping students. Summative assessment involves trying to find out what students can do unaided. This is confounded by some of the instructions for marking given by several examining groups which result in fewer marks being awarded if the student is helped. For example:

Design and carry out an experiment to solve a problem

Zero	Needs continual guidance and still does not reach a solution
1 mark	Can reach a reasoned solution only with continual guidance and reassurance
2 marks	Proceeds effectively after being given guidance
3 marks	Designs and performs an experiment following safe procedures to a reasoned solution

Such instructions make teachers reluctant to help their students because by doing so they reduce the marks that they can properly give. A major problem is that teaching in the laboratory normally takes place with students working in groups of two to four but the examination system requires information about individuals. Many teachers, in trying to be fair to all their students and at the same time not giving them any help, set up special assessment occasions in order to get the information needed for the summative assessment. They feel that they are not able to extract the information they require from the normal process of formative assessment which they are doing all the time and so they adopt the attitudes and procedures of the old external practical examination. This runs counter to the spirit of coursework assessment and to the requirement of most examining groups that teacher assessment should be done as a part of normal teaching. In the National Curriculum, where far more than the assessment of a few practical skills is needed, it will be vital for teachers to break away from the examination syndrome if they are to do any teaching and not to spend most of their time examining.

Three interconnected changes are needed to resolve the conflict. The first is for teachers to look at the process of teacher assessment from a teaching and not an examining point of view. The second is for teachers to have confidence in their own professional judgement. The third, and perhaps the most difficult, is for everyone to increase their trust in the professional judgement of teachers. This means going through a process something like the following:

- Know what is thought to be important, perhaps by having clearly laid-out teaching and learning objectives. The assessment objectives in the GCSE may be suitable but should be renamed.
- Concentrate on the achievement of these objectives.
- Think always about formative assessment.
- Keep records of how the students are getting on.
- When needed, extract from these records the information which an examination board or anyone else may legitimately require.

Confidence among teachers can be increased by gaining a greater awareness of the examining process, of what assessment can and cannot do, and of the inherent unreliability of any measurement, particularly that of the ability of human beings. It can be increased further by a shared knowledge of standards and of methods of assessment. The proposals in the Task Group on Assessment and Testing (TGAT) report (DES 1988a) for the interactive moderation of teacher assessment in the National Curriculum are a great step forward. Mutual awareness of standards and methods of assessment through a process of regular meetings was a feature of the Schools Council Integrated Science Project (SCISP) examination run by the Associated Examining Board, and is a part of the procedures of OCEA (see Willmott *et al* 1987, for more details.)

Increasing the trust in the professional judgement of teachers is the most difficult of these changes because it involves educating everyone from politicians to pupils about what assessment can and cannot do, and about the relationship between assessment and education. It will be a long and difficult task, but one good outcome of the National Curriculum will be that more people will become more directly involved. Governors and parents, for example, will be made more aware of the issues which are involved and will have to think more deeply about what it all means.

The examining groups can play a more direct and immediate part by making clear what their role is, particularly in relation to the moderation of teacher assessments. In the moderating process they are *not* checking on the honesty of teachers, they ought *not* to be examining individual pupils, but they *should* be trying to arrive at common standards among different teachers. Actually they usually check for common standards among different schools, and rely upon individual schools to ensure common standards among the teachers in that school. For this purpose they need to see examples of the kind of work which a school feels merits a particular mark. They need to see examples of '5-ness', '3-ness' and '1-ness', and so do not need to see all the work of any individual pupil nor even the work of any named pupil. If moderation is seen in this way, it will reduce the pressure on teachers who feel they have to save everything they do and justify every decision they make. In principle, if the standard of a school is constant, the examining group does not even need to see the work of the candidates who are being examined that year, nor check every school each year. The samples of '5-ness', '3-ness', and so on, from a school can be spread over several years. However, the standard of an individual teacher may shift with time as experience is gained about what other people are doing. In addition, new teachers join a school from time to time. This means it makes sense to moderate fairly frequently and it makes sense to moderate every school every year in the early stages of a new system.

It would also help if there were fewer detailed instructions issued to try to achieve reliability in an assessment which is inherently unreliable. Instructions, such as spelling out the number of practical tasks which have to be assessed and the details (checklists) of the points to look for in order to award marks, increase the pressures and reduce the room for teacher judgement. What is needed are plenty of exemplars and fewer instructions. In the early stages of the GCSE the examining groups were under pressure from many teachers to give specific instructions about what to do. As experience is growing there are signs that teachers and examiners are beginning to relax and so, perhaps, the pressures will decrease.

It seems that the move from tight controls to more freedom is symptomatic of any change of this nature. We can see the demand from teachers to be told what to do being repeated in the National Curriculum, only this time it is not the GCSE examining groups who can provide the answers. In principle it is the School Examinations and Assessment Council (SEAC) who will do it. In practice, as with the GCSE in fact, it will be the teachers themselves who will provide the answers. The SEAC will have to go to teachers, through its committees, through the teams working on the Standard Assessment Tasks, and so on. This is what the GCSE groups did; and this brings us back to mutual education about standards and to teachers having confidence in their own judgements.

CHAPTER 15

Assessment and evaluation in the science laboratory

Geoffrey J. Giddings, Avi Hofstein and Vincent Lunetta

Introduction

We view the science laboratory as a unique education setting which incorporates a variety of instructional strategies and materials. This complex setting involves the student, the teacher and the learning materials (laboratory manuals, instructional guides, materials and equipment). Together these variables interact to form the learning environment one calls the school science laboratory. This chapter will review the evaluation tools and methods that can be used to assess those variables that are either outcomes of laboratory work or play a key role in determining the pedagogical effectiveness of the science laboratory experience.

Precisely what should be evaluated depends, of course, on the general goals of the laboratory as an educational setting. Laboratory activities have been traditionally used to accomplish a wide variety of goals, cognitive, practical and affective (see, for example, Tamir's discussion in Chapter 2).

Science teachers who genuinely want their students to develop the skills and attitudes that relate to specific goals see to it that these kinds of learning find their way into tests and other evaluation procedures. If these goals are important in a given laboratory session, direct incentives for their development will be provided.

Two decades ago Grobman (1970, p. 192), in the United States, identified succinctly the major problem in assessing laboratory performance:

> With few exceptions, evaluation [of laboratory work] has depended on written testing . . . there has been little testing which requires actual performance in a real situation, or in a simulated situation which approaches reality . . . to determine not whether a student can verbalize a correct response, but whether he can perform an operation, e.g. a laboratory experiment or an analysis of a

complex problem . . . This is an area where testing is difficult and expensive yet since in the long run primary aims of [science] projects generally involve doing something rather than writing about something, this is an area which should not be neglected.

It is not unreasonable to suggest that the laboratory provides unique conditions for the acquisition of these cognitive, practical and affective skills. Hence, the assessment and subsequent evaluation of these laboratory skills are important tasks for the teacher, in terms of both assessing student performance and providing important evaluative feedback.

Evaluating student performance

Irrespective of a teacher's particular goals, student behaviours in the science laboratory can be grouped into four broad phases of activity: *planning and design* (formulating questions, predicting results, formulating hypotheses to be tested, designing experimental procedures); *performance* (conducting the experiment, manipulating materials and equipment, making decisions about investigative technique, observing and reporting data); *analysis and interpretation* (processing data, explaining relationships, developing generalisations, examining the accuracy of data, outlining assumptions and limitations, formulating new questions based on the investigation); and *application* (making predictions about new situations, formulating hypotheses on the basis of investigative results, applying laboratory techniques to new problems). While some of these abilities can be enhanced through other activities, the laboratory clearly offers unique opportunities for their development. Unfortunately assessment and evaluation of student performance in laboratory settings has often failed to reflect these important objectives of laboratory-based learning. Broadly speaking, systems for evaluating students activities in the laboratory can be classified into four main categories: *written evidence*, (written reports or items on paper-and-pencil tests); *laboratory practical examinations*; *audiovisual and computer simulations*; and *continuous observational assessment*.

Written material

Although evaluation of students' written reports is one of the most common assessment methods traditionally used, this is often quite subjective because such variables as neatness, writing skills, volume and degree of completeness can bias evaluation. Moreover, the written report, whether completed inside or outside the laboratory, does not provide direct information about students' skills in manipulating equipment, observing, organizing and performing an investigation creatively and efficiently. Additionally, any assessment derived from an individual's written report must be treated with some caution when it results from group activity, since a student may have played an insignificant role in the conduct of the experiment yet still have provided a high-quality report.

Paper-and-pencil practical test items, like written laboratory reports, also suffer from any close examination of their validity. This is essentially because the test items are at best an indirect measure of the actual performance level or skills acquired in the actual laboratory setting. For example, students can be asked to read a temperature scale or a length from a scale printed on a test, but their ability to do so with real equipment might involve other psychometric skills not measured on the written test.

These concerns are supported by research evidence, which has shown that the correlation between students' achievement in practical tests and their achievement based on written evidence is rather low (Robinson 1969; Tamir 1972; Ben-Zvi *et al.* 1977; Doren 1978; however, note also Toh's findings in Chapter 9). There is a clear need to develop special measures to assess not only what the student reports about activities in the laboratory, but also what the student actually does in the science laboratory.

Practical testing

Many manual and other skills relating to a student's resourcefulness in performing an experiment can be assessed in actual laboratory situations. Manipulative skills, observational abilities and more complex problem-solving and process skills can be assessed in this way.

Tamir (1974) suggests that these kinds of practical examination should be designed to meet a number of quite specific criteria. First, he suggests that students should be confronted by some real and intrinsically valuable problem which is comparatively novel to the examinee. Second, it should be possible to complete the investigation within a reasonable time, with the level of difficulty and the required skills commensurate with the objectives and experiences of the curriculum. Marking is then carried out according to a predetermined key of weighted scores for the skills of manipulation, self-reliance, communication, experimental design and measuring. These types of practical examination utilise systematic observations based on a list of specific criteria, as opposed to an open-ended subjective-type assessment.

The open-ended subjective-type assessment is perhaps best reflected in the formal practical examinations that have been in use by examination boards and in final school examinations in several countries. In these examinations, students are usually examined by external examiners and not their own teachers. Such examinations tend to suffer from several drawbacks. Not the least of these is that different examiners tend to apply different criteria to assess student performance. Also, due to the large number of students being tested, examiners are not able to observe each student systematically, so that assessments tend to be based on the actual results of experiments and on written reports. Finally, there are obvious restrictions on the scope and hence validity of the experiments which can be administered to students in a limited time-frame. Obviously these limitations do not apply when practical assessment is used by individual teachers in their own classrooms.

However, teachers in general do not appear to be greatly attracted to this type of formal practical test, due perhaps to the implementation problems which obviously affect the reliability and validity of such assessments. Practical tests can also have undesirable side-effects on a teacher's choice of experiments throughout the year; teachers tending to limit their choice of experiment to those highly related to the type of experiment usually occurring on the practical test. On the other hand, Bryce and Robertson (1985) reported the successful use of the 'stations' technique (pupils circulating around a number of small experiments or practical tasks, usually on a timed basis) in the national implementation of practical assessment for certification purposes. In addition, Garnett and O'Loughlin (1989) report on a successful strategy for implementing laboratory testing at upper secondary and introductory tertiary-level chemistry classes.

Audiovisual and computer simulations

The use of well-designed audiovisual media and computer simulations can make science learning more appealing, interesting and relevant to the student, as well as providing an alternative strategy for the assessment of laboratory skills and abilities (see also Chapter 12 in this book). Computers and audiovisual media have become a widely used strategy in many levels of science instruction. Modern educational technology has developed an enormous range of methods, approaches and techniques designed to improve instruction by facilitating easier and more effective science learning on the part of the student.

The computer as a laboratory partner and a source for obtaining feedback on students was carefully examined in a series of studies conducted at Berkely (Nachmias and Linn, 1987). They reported on microcomputer-based laboratories (MBL) in which the computer was interfaced with traditional laboratory apparatus to collect and graphically display data. Probes were interfaced with the computer to help students measure, record and graph quantities like force, light, heart rate and temperature. One of the most promising features of the use of microcomputers in a laboratory setting is the powerful provision for immediate feedback. While performing the experiment the results are shown clearly on the screen by a table and/or a graph, with the students then being able to analyse the data and make their own interpretations. Students interact with the computer dynamically and the resulting pattern of interaction can be recorded and thus provide continuous feedback on the thinking patterns of each student.

Oliver and Roberts (1974) describe how the introduction of a visual element into assessment procedures, by the combination of filmed sequences in conjunction with conventional multiple-choice tests, extends the range of objectives, hence of behaviours, that can be assessed.

Continuous observational assessment methods

Because each of the assessment and evaluation systems discussed has serious limitations regarding the depth and breadth of skills that can reasonably be measured, more-continuous systems of assessment and observation have been developed. For specific details of assessment models, marking schemes, checklists and rating scales, see Giddings and Hofstein (1980), Ganiel and Hofstein (1982), and Bryce and Robertson (1985). Most teachers would see continuous assessment tasks as a natural part of their role as teachers and as simply an extension of the ongoing monitoring that all teachers carry out in their day-to-day teaching.

Continuous assessment on several occasions throughout the year is necessary to cover adequately the variety of tasks, skills and techniques which comprise a total programme of practical work. It also tends to reduce the incidence of chance success or failure by students. Importantly, with a greater involvement in the continuous assessment of practical skills, the teacher is likely to develop a greater awareness of the scope and objectives of each segment of laboratory work and to identify student strengths that otherwise might not have been reflected in more conventional assessments. Perhaps the most powerful argument in its favour is that if teachers wish to encourage the development of practical skills, then any assessment system must be seen by the students to reflect this aim.

Using continuous observational assessment methods, the teacher unobtrusively observes and rates each student during normal laboratory activities. Observations can be recorded over an extended period of time or during a single laboratory activity. The basic principles of continuous assessment include the following:

1 Teachers should inform their students at the beginning of the course that their practical work is to be assessed in a continuous manner over the whole course. Details should be given regarding the abilities and skills that are to be assessed. Fears have been expressed that the position of the teacher as an assessor may affect adversely the close relationship between student and teacher. Experience suggests that no such concern need exist. Teachers are normally involved in the assessment aspect of a student's work as part of their everyday teaching.

2 As far as possible, the teacher should make assessments during a normal practical class and make their assessment procedure as unobtrusive as possible. It is not necessary to assess all students on the same day or on the same experiment. Also, any single experiment is unlikely to provide the opportunity to give a complete assessment.

3 There are three main methods by which marks can be allocated to a particular assessment and teachers will probably find that they will have to use all of them at some stage of the program:

(i) *A specific mark scheme.* This will be most useful when marking any written or oral evidence of an observation or an interpretation. It will also be useful for marking for planning or accuracy.

(ii) *Marking by impression on a single occasion.* This will be useful for assessing evidence that is less precise than in (i). For example a teacher may wish to assess a student's ability to handle an unfamiliar piece of apparatus by using one of the suggested rating scales on a number of occasions during a laboratory session.

(iii) *Marking by impression over a period of time.* This strategy will be appropriate when attempting to measure attitudinal variables, although some of the less precise aspects of psychomotor skills development may also be better assessed periodically rather than in single experiments.

Teachers also can distribute skills checklists to their students and ask them to rate themselves after each laboratory session or at the completion of a unit of work.

There appear to be clear benefits accruing from the use of observational checklists in many areas of practical science assessment, although it should be noted that detailed checklists can be unsuitable instruments for the assessment of certain aspects of laboratory acitivity. Further, it is obviously important to encapsulate in the total assessment programme, indicators of general performance, particularly in large-scale practical projects. Woolnough (1986, p. 195) however, alerts us to some of the pitfalls in concentrating too much on specific assessments when he argues that the Assessment of Performance Unit (APU) has

shown that students, even those who do not do well on tests of specific practical skills, will often perform very competently on a problem to which they can relate. Such insights have caused many of us to recoil from going down the road of tightly prescribed behavioural objectives, and to emphasize and encourage the essence of genuine scientific activity through the assessment of whole investigations.

Assessment of student attitudes and student perceptions of the laboratory environment

The laboratory is an effective environment for enhancing student attitudes to and interest in science learning. Students' attitudes to science laboratory activities, to science as it is practised in the laboratory, and their perceptions of the science laboratory as a learning environment, are important areas to be targeted by teachers in their assessment and monitoring of the science laboratory experience.

Attitudes to the laboratory setting

The development of a positive attitude to science and the scientific enterprise is among the major aims of science teaching. However, there is often some confusion about what meaning should be placed on the term 'attitude to science'. Klopfer (1971), however, has alleviated many of the problems associated with the multiple meanings attached to the term by providing a comprehensive classification scheme for science education aims in the attitudinal area in which six conceptually different categories of attitudinal aim are distinguished. These six categories define distinctions between: attitude to science and scientists; attitude to inquiry; the adoption of scientific attitudes like curiosity and open-mindedness; the enjoyment of science learning experiences; interest in science apart from learning experiences; and interest in a career in science.

In one test, the *Test of Science-Related Attitudes* (TOSRA), Fraser (1978) provides coverage of each of these distinct categories indentified by Klopfer. Each scale in the TOSRA contains ten items in five-point Likert response format. In another test, Hofstein *et al.* (1976) developed and validated a measure to assess students' interest in and attitude to high-school chemistry laboratory work. Analysis of students' responses using a simple factor analysis procedure revealed that a student's attitude towards the science laboratory is not unidimensional. The following attitudinal dimensions were obtained: 'learning in science laboratory'; 'amount of laboratory work'; 'value of laboratory work'. It was found that the measure was sensitive to gender differences as well as to grade differences. For example, it was observed that chemistry students in the twelfth grade (age 17) found laboratory work less stimulating compared to their eleventh and tenth grade counterparts. This information may well be important for both curriculum developers and school science teachers in the planning of activities and experiences for students at different levels.

Attitude instruments such as these can be used by teachers and curriculum evaluators to monitor student progress towards achieving attitudinal aims, particularly those that are derived from the laboratory experiences. Although it is possible to use an attitude test like the TOSRA for assessing the progress of individual students, it is likely to be most useful for examining the performance of groups or classes of students. Furthermore, as well as providing information about attitudes at a particular time, TOSRA can be used as a pre-test and a post-test (perhaps over a school term or year) to obtain information about attitude changes.

Perceptions of the laboratory environment

Because students spend a huge amount of time at school, they have a large stake in what happens to them there, and their reactions to and perceptions of their class experiences are significant. Here we consider one approach to conceptualising, assessing and investigating what happens to students during their schooling. The main focus is upon students' and teachers' perceptions of important social and psychological aspects of the learning environments of science laboratory classrooms.

Since the early 1970s information has been gathered in many countries concerning the classroom learning environment and its relationship to outcomes such as student achievement and attitudes (Fraser and Walberg 1989). A sizeable proportion of this research has involved science classes specifically. For example, Fisher and Fraser's (1982) study involving 116 Australian science classes established sizeable associations between several inquiry skills and science-related attitudes and classroom environment dimensions. This approach finds strong support in Hofstein and Lunetta's (1982, p. 212) review of science laboratory teaching when they advocate that science

> laboratory activities have the potential to enchance constructive social relationships as well as positive attitudes [to science] and cognitive growth. The cooperative team effort required by many laboratory activities may promote positive social interactions involving cohesiveness, task orientation, goal direction, democracy, satisfaction and other factors – all of which can be measured by classroom environment measures.

In light of the success of this approach, a new classroom environment instrument has been developed in Australia which specifically aims at measuring student perceptions of science laboratory environments (Giddings and Fraser, 1989). This instrument, the Science Laboratory Environment Inventory (SLEI), assesses eight dimensions (see Table 15.1) of a student's actual and preferred environment in a science laboratory class at the secondary and higher education levels. This has the dual advantage of characterising the class through the eyes of the actual participants and capturing information which the teacher could miss or consider unimportant. It provides the classroom teacher and the researcher with an additional evaluative tool to assess important dimensions of the science laboratory experience, and to modify it appropriately.

Assessing and evaluating laboratory instructional materials

The laboratory is the key setting where students actually carry out the process of scientific inquiry. In doing so, the laboratory manual and other instructional materials play a major role for most teachers and students in defining the goals and procedures of the science laboratory activities. Whatever these goals and procedures, it is important to be able critically to analyse and hence decide whether any given laboratory work or project will contribute to achieving them. In order to do so, one needs a method of analysing and assessing laboratory manuals and other instructional material.

There are vast differences in teaching strategy which differentiate one kind of laboratory activity from another and clearly affect learning outcomes. Some laboratory activities are organised in a didactic manner, whereby students gather data to verify relationships and laws that have been outlined in a textbook or by the teacher in the class. Other laboratory activities precede the formal introduction of a

Table 15.1 Descriptive information for each scale in the Science Laboratory Environment Inventory (SLEI)

Scale name	Description	Sample item
Teacher supportiveness	Extent to which the teacher/instructor is helpful and shows concern for all students.	The teacher is concerned about students' safety during laboratory sessions. (+)
Involvement	Extent to which students participate actively and attentively in laboratory activities and discussions.	During laboratory group students leave it to their partners to do all the work. (−)
Student laboratory cohesiveness	Extent to which students know, help and are supportive of one another.	Students in this class get along well as a group. (−)
Open-endedness	Extent to which the laboratory activities emphasise an open-ended, divergent, individualised approach to experimentation.	We know the results that we are supposed to get before commencing a laboratory activity. (−)
Integration	Extent to which the laboratory activities are integrated with non-laboratory and theory classes.	We use the theory from our regular science class sessions during laboratory activities (+)
Organisation	Extent to which the laboratory activities are clearly defined and well organised.	There is confusion during laboratory classes. (−)
Rule clarity	Extent to which behaviour in the laboratory is guided by formal rules.	There is a recognised way of doing things safely in laboratory. (+)
Material environment	Extent to which the laboratory equipment and materials are adequate.	The laboratory is too crowded when we are doing experiments. (−)

Items designated (+) are scored 1, 2, 3, 4 and 5, respectively for the responses Almost never, Seldom, Sometimes, Often and Very often. Items designated (−) are scored in the reverse manner.

topic and involve students in gathering information about materials or phenomena from which they will subsequently infer relationships and make generalisations.

Additionally, laboratory activities can vary greatly from one another with respect to the amount of guidance provided to students. Some laboratory investigations are highly structured, with students following detailed recipe-type instructions, while others are much more 'open-ended' and involve students in more of the elements of an experiment, such as the planning and design. Further, while some laboratory activities emphasise manipulation of materials, other emphasise observational skills, the interpretation of data, or the application of known procedures to new problems. While it is highly probable that such differences among laboratory activities affect learning outcomes, it is even more likely that interaction effects will also exist between the style of the laboratory and the prior learning and logical development of individual students.

A number of methods of analysing the learning experiences provided to students through laboratory manuals have been suggested in the literature. Recognising the need to examine the nature and quality of written laboratory manuals, Fuhrman *et al.* (1978) have developed a task analysis inventory called the laboratory Analysis Inventory (LAI). They identified 24 skills that could be clustered under the following four instructional modes: planning and design; performance; analysis and interpretation; and application. This inventory was found to be a useful instrument in analysing laboratory manuals used in different science curricula. Since the task categories include actual behaviours required to perform prescribed laboratory work, it can be used as an evaluative tool by teachers who attempt to tailor their science curriculum and laboratory activities to their unique classroom setting.

Kuehl (1984), in the United States, developed a method to analyse laboratory instructional materials in the context of physics education. A task analysis instrument was designed to help high-school teachers select a laboratory manual that would complement their personal teaching style. There are a large number of other such tools and checklists designed for this purpose. It is suggested that the use of analysis systems like the two described can provide both teachers and researchers with sensitive methods of gauging the values and objectives implicit in their instructional materials and how their students interact with them.

Evaluating teacher behaviour in the laboratory

The final focus of this chapter is on the teacher's own behaviour in the laboratory setting. For example, an experiment can be open-ended and inductive when taught by one teacher, but deductive when taught by another. Teaching style does make a difference. Since the accepted role of the laboratory has been brought into question, it is important to be able to assess and evaluate not only what the students are doing in the laboratory but also the teachers, and, in particular, the way in which teachers are translating the aims and objectives of curriculum writers and developers into action in the laboratory.

Eggleston *et al* (1976), in their UK study, found that teaching style tends to be consistent no matter what form of activity takes place. That is, an experiment can be open-ended and inductive when taught by one teacher but deductive when taught by another, depending on the teachers' individual styles. Additionally, they found that deductive-orientated teachers tended to teach practical work authoritatively, while more inquiry-orientated teachers tended to teach an investigative method of learning.

In order to obtain more objective information about teaching practice and about the interactions among teachers, curriculum resources and students, there is a need to develop objective and reliable methods of analysing teachers' behaviours in the science laboratory. An important attempt to develop a systematic science classroom analysis of what actually happens in the science laboratory was made by Penick *et al*. (1976), who developed their *Science Laboratory Interaction Category (SLIC) – Student* instrument. Shortly after, Shymansky *et al*. (1976) developed the *SLIC – Teacher* version of the instrument. Using these two instruments, it has been possible to obtain important information about the kind of teaching and learning that takes place in the science laboratory.

Another example comes from Israel. Tamir (1977) used the classroom observation schedule originally developed by Smith (1971) to observe students conducting experiments in the biology laboratory. The instrument provides a record of teachers' and students' behaviour during the pre-laboratory and actual laboratory phases and post-laboratory activities. Tamir found that in Israel high-school biology was taught, by and large, using inquiry techniques, whereas at university level, the laboratory work reflected a more traditional recipe-style approach.

Recent applications of laboratory observation instruments and classroom environment instruments have supported the effectiveness of such strategies and attested to the potential usefulness of employing such instruments. They can provide science teachers with meaningful information and feedback about problem areas and a tangible basis for guiding further improvement in the use of student science laboratories.

PART VII

Epilogue

CHAPTER 16

Practical science as a holistic activity

Brian E. Woolnough

It is not appropriate, at the end of this collection of papers, to attempt to produce a tidy summary. But it is the privilege of the instigator and editor of the book to have the last word in the light of what has been written, and this I intend to express.

The questions revisited

We started the book with a series of questions which have, to a greater or lesser extent, been addressed by the authors. Perhaps the most significant questions have been the ones that have not been resolved, in some cases almost ignored. The sound of the 'dog that did not bark' can be most significant in indicating areas that need further consideration.

Justifications for doing practical work in science have been well argued throughout this book. The conviction of all the authors that practical work is vital, indeed central, to science teaching is self-evident. The main difference lies between those teachers who see practical work being done primarily to enhance the students' understanding of the theories of science and those who see it as being done primarily to develop their ability to do practical problem-solving science. Clearly, different types of practical need to be done to meet these two goals.

Where the aim is to develop theoretical insights, the model for practical work has often been seen as tightly structured, to ensure that 'the right theory' has been enhanced. There is still much practice of standard exercises, with the student being expected to follow 'cookbook' instructions. The evidence is that such practical work does little to enhance students' understanding of the concepts of science and nothing to enhance their appreciation of the methods of science. Gunstone speaks convincingly of the difficulty of using practical work as a way of reconstructing students' knowledge, as the clutter and complexity of the equipment become distracting and makes it an end in itself. The reconstruction will certainly not happen automatically. Both Gunstone, with his POEs, and White, through his episodes, have stressed the need to force the students to think, consciously, about what they are doing and explicitly to make links for them.

Where the aim has been to teach students how to work as scientists, the emphasis has often been on exercises to develop scientific skills rather than on complete investigations to develop full scientific capability. Millar has demonstrated that limitations and invalidity of such a process approach. It is evident that even when concentrating on the skills and processes of science, it is only as they are developed in the context of doing science that they acquire any scientific validity. Practical investigation must be at the heart of science teaching. Gott and Mashiter and Kimbell emphasise the importance of such being done in the context of complete tasks. Such investigations may be short or extended, related to pure scientific relationships or more technological problems, but each will have the same structure. They will start with a problem, which the student will then take, turn into a manageable task and plan a scheme of work. That plan will then be executed with apparatus selected or constructed, measurements taken, observations made, and results recorded and interpreted. At each stage of the investigation the planning will be refined as feedback is obtained. The investigation and the conclusions will be reported in an appropriate manner for communication. The Assessment of Performance Unit (APU) established a problem-solving model for science along these lines (see Figure 14.1), which is similar in iterative process, if not in context, to the way Kimbell describes the technological approach. Such investigations not only parallel the way in which scientists work, and thus introduce students to this important cultural as well as useful approach, but also are highly motivating and thus effective. The inclusion of the planning stage in practical work not only reflects good scientific practice but also transforms the activity from being the teacher's experiment to being the student's own. The student thus takes personal responsibility for the experiment and acquires ownership of it. Active, generative learning can then take place. Toh has shown, encouragingly, that the explicit teaching of problem-solving strategies can, in part at least, develop the student's competence in tackling scientific investigations. We need to note well the evidence provided by Murphy that in some cultures girls and boys tackle problems in different ways and will respond differently to different tasks and teaching strategies.

More work still needs to be done on the relationship between theory and practical work. We know that much practical work, because of the vast amount of experimental clutter and information overload, has been ineffective in increasing theoretical understanding. But it can, if it is appropriately structured, and the appropriate links are provided, enhance a half-formed understanding and build up the all-important sense of interest, achievement and self-confidence. The last phrase of the old maxim 'I hear and I forget; I see and I remember; I do and I understand' needs to be changed not to 'I do and I get even more confused' but 'I do and I believe' or even 'I do and I build up confidence that I can tackle the next problem successfully'. Clearly different practical experiences will be needed for different purposes; Gunstone's shorter and more conceptually focused practical qualitative tasks, White's episodes, Allsop's environmentally based tasks, all aim to strengthen this linkage through personal commitment.

The inverse relationship, as to how much the understanding of theory helps to solve practical problems, is complicated by the fact that so much of the effective knowledge applied is tacit and not explicit. We find that students know far more than they can tell. Their ability to apply procedures and knowledge which they have gained through personal experience, yet which they could not articulate, is considerable and vitally important. We ignore such tacit knowledge at our peril, and must guard against undervaluing it because we cannot measure and assess it. The other issues so often ignored relate to the vital importance of social interactions (see Solomon's contribution) and affective factors such as motivation. Students may fail not because they *cannot* but because they *do not want to*. A student who wants to succeed on a scientific task, and has the self-confidence to tackle it, may reveal ways of working and a grasp of the underlying theory which was unsuspected and could never be made explicit. Scientists utilise craft skills, demonstrating a confident feel for materials and appropriate procedures gained through personal experience and success. If we are to capitalise on this we must plan students' practical science more with a view to the experience they will gain and the success they will achieve than for the techniques and the knowledge they will learn explicitly.

Perhaps the two most contentious aspects relate to transferability and progression. Though there are many who would claim, explicitly or by inference, that taught practical skills and processes are transferable, there is little research evidence to confirm that such transferability takes place other than over a very restricted range of tasks, though Toh has shown where it can occur. Such limited transferability may of itself be important and worth teaching. But Miller, Gott and Mashiter and Gunstone warn us that we should beware of claiming too much, especially that some of the general processes such as observation are the same as scientific observation when tackling a scientific problem. It is easy to analyse performance on a scientific task in terms of the knowledge and procedures involved. At the end of the day, however, it may well be that it is the missing component, the affective factors of motivation and self-confidence, which are the only ones to have any high degree of transferability. And if that is so, we must redirect the criteria for selecting appropriate practical science. Not that it should cover the right content or develop the right skills but that it should provide satisfying experience for the student leading to success. Then the transferability will depend as much on the motivation generated by the new task as the skills and knowledge brought from previous tasks.

As to the concept of progression, that the students progress in their scientific skills and processes along some uniform path, I suspect that we may be searching in vain for any definable form of predetermined progress, though Gott and Mashiter and Fairbrother have given some useful guidelines. We know, in retrospect, that some students do better than others in tackling a scientific problem, we know that a particular student has improved over a period of time and we know what aspects of the students' work we need to encourage. We know also that some tasks are inherently more difficult than others. But we cannot predict, in advance, the path along which any particular student will develop. Furthermore, that development may well be demonstrated in quite different ways on different tasks. I suspect that progress in scientific learning is more like a roller-coaster ride than a walk along a straight road, and even then the analogy falls down because of the predetermined rails of a roller-coaster. It is interesting to note that while all science educators wish to encourage progression in the students' learning it is the English in particular who have become obsessed with the need to define it. This is a result of the move towards criterion referencing and an elaborate assessment scheme designed to deliver a national curriculum in terms of ten definable steps from the age of 5 to 16 (DES 1988a). I have argued that such an assessment structure is not only administratively unworkable but also educationally invalid (Woolnough 1989b). Other countries would do well to beware of following such a pattern based, as I believe it is, on a spurious concept of progression. We can make the tasks progressively more difficult, and we can then define progression in terms of the task successfully tackled. But to search for predetermined progression in student ability to do practical science is likely to prove as elusive as the search for the holy grail.

Fundamental questions must still remain about the effect, as well as the methods, of assessment in science. Of course, a teacher will want to assess a student's progress so that difficulties can be overcome, and progress reported. It is important for the student, as much as for the teacher, to be clear about what aspects are being encouraged in doing science and thus, thereafter, assessed. Such clarification has been necessary and beneficial. But dangers have arisen when assessment has moved from light, sensitive, informal judgement based on the teacher's informed professional judgement to tightly prescribed assessment procedures which attempt to measure reliably specific, predetermined scientific skills. Such atomistic assessment, moving away from holistic assessment of the overall performance on the scientific investigation, can easily lead to a fragmentation of scientific activity into trivial tasks or a spuriously precise criterion description of broad processes. Fairbrother and Kimbell have discussed these issues. I believe that scientists have much to learn from the technologist's more holistic approach here. This is not the place to discuss this danger further, or the possible alternative approaches that are available, I have done this elsewhere (Woolnough 1988b; Woolnough and Toh 1990). Suffice to stress here the need to ensure that the form of assessment used, which will inevitably determine the way that the science is taught, does not have a detrimental effect on the student's experience of science. If we are to encourage a holistic approach to practical science, by stressing the importance of students' doing complete investigations, we must ensure that we assess it holistically, too. An assessment framework which reduces the whole to a series of measurable parts can easily restrict and destroy the promotion of genuine scientific activity. While recognising that assessment must be at the heart of good science teaching, it must be there as a subservient not a determinent factor. The practice of good science can be encouraged more by the carrot of example and exhortation than the stick of assessment.

Reductionism or holism?

As we move towards a positive definition and reaffirmation of practical science as a holistic activity, we need to be clear about our meaning of the increasingly overused word 'holistic'.

There has been a tendency since the 1970s to a form of reductionism in science education, as in other aspects of life. With the welcome move to clarifying our goals, we have followed Bloom (1956) in defining sets of behavioural and assessable objectives. But in so doing we have reduced scientific activity to a set of measurable but not very important parts. By insisting on assessment, we have omitted those factors that cannot be measured. By insisting on accountability and reliability we have stressed the parts at the expense of the whole. We have, furthermore, implied that we can create the whole by combining a series of component parts. Yet producing scientists is more like growing a plant than building a house from bricks, more like conceiving and nurturing a baby than building a robot from spare parts.

In seeking to develop in our students the ability to become, or to appreciate, problem-solving scientists we need to remember three principles. First, the whole does not equal the sum of the parts. Second, the whole is greater than the sum of the parts. Third, the whole is altogether more powerful than the sum of the parts. Then we must encourage opportunities in our teaching for students to develop their ability and confidence in doing science by doing whole scientific investigations.

Holistic approaches have been used to describe many differing approaches, even including homoeopathic medicine. In science education the holistic approach has variously included an embracing of skills and processes into the whole ability to solve practical problems, an embracing of knowledge and skills into the whole problem-solving process, and an embracing of knowledge, skills and attitudes into the whole. It is in this latter, all-embracing, sense that I will discuss the way forward, for I believe that we have traditionally ignored those affective aspects of students' learning, aspects of motivation, commitment, self-confidence, and student satisfaction, which hold the key to student success and fulfilment in practical science.

Practical science as a holistic activity

I have found it helpful to think of practical activity in school science as consisting of three types – investigations, exercises and experiences. At the heart of scientific activity must be the practical *investigation*. Whether such investigation last for a few minutes or a few weeks, whether it concerns a scientific relationship or a technological problem, the process of planning, performing, interpreting and communicating, with its continual modification through feedback, is fundamental to the way in which scientists work. Other practical work may lead up to this most complete form. When it is necessary to develop a particular skill or to become familiar with a particular piece of apparatus a practical *exercise* may be needed, though even here it is appropriate to incorporate that exercise into a genuine scientific activity rather than attempting to develop that skill out of context. Finally, practical *experience* is designed quite specifically to give the student a feel for the phenomena under investigation, to build up personal experience and tacit knowledge which will form the basis for subsequent action and understanding as links are formed. The practical investigation, with its personal commitment to the problem and the continual interaction between hand and mind, will allow ideas to be tested and refined and experience to shape the explicit and the tacit understanding of the student.

Hodgkin (1985) developed an oscillating model for learning, influenced, I suspect, by his mountaineering background. He stresses that personal commitment and challenge, springing from a supportive and stimulating environment, are essential for personal growth and learning. He categorises three types of activity: play to build up interest and basic familiarity (in the foothills); practice to build up basic competences and confidence (on the training slopes); and exploring to stretch, challenge and extend the frontiers (on the mountain peaks). He argues that a learner needs to be continually moving from one type of activity to another, to enjoy a wide range of experience – first playing in the foothills to stimulate interest, then practising to build up competence in the required skills before facing a fresh and extending challenge in exploring new regions, and returning with renewed self-confidence to discover a new area in which to play. The affective aspects of interest, enjoyment, commitment, perseverence, self-confidence and challenge are fundamental throughout. The student needs to be given freedom to play, to practise and to explore, the teacher's main tasks being to create the space, sustain the dialogue and introduce structures, problems and evaluations. The parallels in practical science are clear. The experiences parallel the play, the exercises the practice and the investigations the exploring. I would contend that the affective aspects are equally fundamental to learning science, and that a science course which is too tightly structured will seriously restrict individual growth.

- Play > Practical Experiences to build a feel for the phenomena and an interest in the area
- Practice > Practical Exercises to develop competence in specific skills and techniques
- Exploring > Practical Investigations to aquire stimulation, confidence and the ability to work as a problem-solving scientist

Whether the sequence should be that of an oscillating cycle or a spiral is not important; I suspect the latter portrays more easily the image of growth. If we build a holistic model we will need to include a range of student attributes, far wider than the traditional explicit knowledge and procedural skills. We will need to start with the range of preconceptions, skills, expertise and attitudes that the student brings into the science lesson, gained from a dozen or so years of experience in life outside school. This experience will have built up a range of tacit knowledge, of the properties and behaviour of things, as will all of the subsequent experience in school. Students will also have at their disposal a range of skills and ways of solving problems, some of which they have been taught, many of which they have picked up. In school the student's insight and knowledge will grow, through explicitly focused discussion, exposition and consolidation, and this also will be increasingly available for use in tackling problems. But the final group of attributes that a student brings to tackle a scientific task involves attitude such as commitment, determination and, above all, self-confidence. As students gain experience and confidence by tackling scientific investigations, all aspects will develop in an increasing, but unpredictable, spiral. The very act of representing such a model with a static diagram can imply the fragmentation of the whole into component parts. At the risk of such misrepresentation Figure 16.1 highlights the various attributes.

All of these attributes are involved when a student comes to a problem, interacting in a holistic way. Practical science provides the opportunity to develop them all, together, not in a reductionist way of trying to develop each in isolation of the others; certainly not by concentrating on the knowledge and skills elements only and trying to build up from them; not by insisting that each component is separated

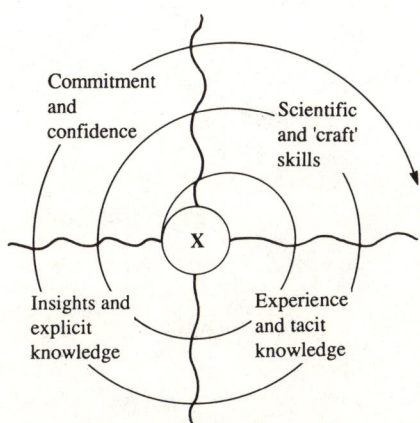

X = Extra-curricular experience, skills,
attitudes and preconceptions brought
by the student into the lessons.

Figure 16.1 Student attributes growing through practical science as a holistic activity

out to be assessed reliably to see how well it has progressed since it was last measured; but by giving students the opportunity to play, to practise and to explore in a safe but stimulating environment as they investigate scientific tasks in the laboratory and the local environment. If we can leave our students with a sense of self-confidence in their ability to tackle scientific problems and have stimulated them by the fun and challenges of science, we will have equipped them with vision and a pair of stout walking boots well prepared to deal with the next unexpected challenge.

References

Ahmed, R. (1977). 'Science education in the rural environment'. In *Science Education in the Asian Region*. Bangkok, Unesco, p. 190.

Alvord, D.J. (1972). 'Achievement and attitude', *The Science Teacher*, 39, 36–8.

American Association for the Advancement of Science (1967). *Science – A Process Approach*. Washington DC, Ginn & Co.

American Association for the Advancement of Science (1989). *Science for All Americans: A Project 2061 Report on Literacy Goals in Science, Mathematics, and Technology*. Washington DC, AAAS.

Anderson, H. (1968). 'The teaching in the laboratory'. In H. Anderson (ed.), *Readings in Science Education*. New York, Macmillan, pp. 28–30.

Anderson, J. (1984). *Computing in Schools*. Australian Education Review, no. 21. Australian Council for Educational Research.

Assessment of Performance Unit (1984). *Science in Schools: Age 15 no. 2*. London, DES.

Assessment of Performance Unit (1987). *Design and Technological Activity: A Framework for Assessment*. London, HMSO.

Assessment of Performance Unit (1989a). *Science at Age 13*. London, HMSO.

Assessment of Performance Unit (1989b). 'A response from the APU team to the Interim Report of the National Curriculum Design and Technology Working Group', *DESTECH Journal*, Series 2, January.

Atkinson, E.P. (1980). 'Instruction–memory–performance: the influence of practical work in science on memory and performance'. Unpublished doctoral thesis, Melbourne, Monash University.

Ausubel, D.P. (1968). *Educational Psychology: A Cognitive View*. New York, Holt, Rinehart and Winston.

Baird, J.R. (1986). 'Improving learning through enhanced metacognition: a classroom study', *European Journal of Science Education*, 8, 263–82.

Baird, J.R. and Mitchell, I.J. (eds) (1986). *Improving the Quality of Teaching and Learning: An Australian Case Study – The Peel Project*. Melbourne, Faculty of Education, Monash University.

Baird, J.R. and White, R.T. (1982a). 'A case study of learning styles in biology', *European Journal of Science Education*, 4, 325–37.

Baird, J.R. and White, R.T. (1982b). 'Promoting self-control of learning', *Instructional Science*, 11, 227–47.

Barnes, D. (1976). *From Communication to Curriculum*. Harmondsworth, Penguin.

Bates, G.R. (1978). 'The role of the laboratory in secondary school science program' in Rowe, M.B. (ed.), *What Research Says to the Science Teacher*, Vol. 1. Washington DC, National Science Teachers Association.

Beatty, J.W. and Woolnough, B.E. (1982). 'Practical work in 11–13 science', *British Educational Research Journal*, 8, 23–30.

Bentley, D. and Watts, M. (eds) (1989). *Learning and Teaching in School Science*. Milton Keynes, Open University Press.

Ben-Zvi, R., Hofstein, A., Samuel, D. and Kempa, R.F. (1976a). 'The effectiveness of filmed experiments in high school chemical education', *Journal of Chemical Education*, 53, 518–20.

Ben-Zvi, R., Hofstein, A., Samuel, D. and Kempa, R.F. (1976b). 'The attitude of high school students towards the use of filmed experiments', *Journal of Chemical Education*, 53, 575–7.

Ben-Zvi, R., Hofstein, A., Samuel, D. and Kempa, R.F. (1977). 'Modes of instruction in high school chemistry', *Journal of Research in Science Teaching*, 14, 433–9.

Bloom, B.S. (1956). *Taxonomy of Educational Objectives. 1, the Cognitive Domain*. London, Longman.

Bloom, B.S. (1976). *Human Characteristics and School Learning*. New York, McGraw-Hill.

Blosser, P. (1981). *A Critical Review of the Role of Laboratory in Science Teaching*. Science Education Information Report. Columbus, Ohio, Center for Science and Mathematics Education, Ohio State University.

Bork, A.M. and Robson, J.A. (1972). 'A computer simulation for the study of waves', *American Journal of Physics*, 40, 1228–94.

Boulanger, F.D. (1981). 'Ability and science learning: a quantitative analysis', *Journal of Research in Science Teaching*, 18, 113–21.

Bourque, D.R. and Carlson, G.R. (1987). 'Hands-on versus computer simulation methods in chemistry', *Journal of Chemical Education*, 64, 232–3.

Brandwein, P.F. (1981). *Memorandum: On Renewing Schooling and Education*. New York, Harcourt Brace Jovanovich.

Brandwein, P.F., Morholt, A. and Abeles, S. (1988). 'Apprenticeship to well-ordered empiricism: teaching arts of investigation in science' in Brandwein, P.E. and Passow, A.H., *Gifted Young in Science*. Washington DC, National Science Teachers Association, pp. 273–306.

Brilliant, A. (1979). *I May Not Be Totally Perfect, But Parts of Me Are Excellent*. Santa Barbara, CA, Woodbridge Press.

Bruner, J.S. (1966). 'Some elements of discovery' in Shulman, L.S. and Keisler, E.R. (eds), *Learning by Discovery: A Critical Appraisal*. Chicago, Rand MCNally.

Bryce, T.G.K., McCall, J., MacGregor, J., Robertson, I.J. and Weston, R.A.J. (1987). *TAPS Report of the Project Phase 2 1983–86*. Edinburgh, Jordan Hill College of Education.

Bryce, T.G.K., McCall, J., MacGregor, J., Weston, R.A.J. and Robertson, I.J. (1983). *Techniques for the Assessment of Practical Skills in Foundation Science*. London, Heinemann.

Bryce, T.G. and Robertson, I.J. (1985). 'What can they do? A review of practical assessment in science', *Studies in Science Education*, 12, 1–24.

Carey, S., Evans, R., Honda, M., Jay, E. and Unger, C. (1988). '"An experiment is when you try it and see if it works": a study of junior high school students' understanding of the construction of scientific knowledge'. Research Report, Educational Technology Center, Graduate School of Education, Harvard University, Cambridge, MA.

Caven, C.S. and Logowski, J.J. (1978). 'Effects of computer simulated or laboratory experiments and student aptitude on achievement and time in a college general chemistry laboratory course', *Journal of Research in Science Teaching*, 15, 455–63.

Chalmers, A.F. (1982). *What Is This Thing Called Science?*, 2nd ed. Milton Keynes, Open University Press.

Champagne, A.B., Gunstone, R.F. and Klopfer, L.E. (1985). 'Effecting changes in cognitive structures among physics students' in West, L.H.T. and Pines, A.L. (eds), *Cognitive Structure and Conceptual Change*. Orlando, FL, Academic Press, pp. 163–87.

Champagne, A.B., Klopfer, L.E. and Gunstone, R.F. (1982). 'Cognitive research and the design of science instruction, *Educational Psychologist*, 17, 31–53.

Charlesworth, M., Farrall, L., Stokes, T. and Turnbull, D. (1989). *Life among the Scientists*. Melbourne, Oxford University Press.

Chodorov, N. (1978). *The Reproduction of Mothering*. Berkeley, University of California Press.

Choi, B. and Gennaro, E. (1987). 'The effectiveness of using computer simulated experiments on junior high students' understanding of the Volume Displacement Concept', *Journal of Research in Science Teaching*, 24, 539–52.

Clement, J. (1982). 'Student preconceptions in introductory mechanics, *American Journal of Physics*, 50, 66–71.

Cole, M.J.A. (1975). 'Science teaching and science curriculum development in a supposedly non-scientific culture, *West African Journal of Education*, 19, 308–13.

Computers in the Curriculum Project (1983). *Acoustics*. London, Longman.

de Bono, E. (1970). *Lateral Thinking*. London, Ward Lock Educational.

Department of Education and Science (1981). *Science in Schools – Age 11: Report No. 1*. London, HMSO.

Department of Education and Science (1985a). *Science at Age 15. Science Report for Teachers No. 5*. London, HMSO.

Department of Education and Science (1985b). *Science 5–16: A Statement of Policy*. London, HMSO.

Department of Education and Science (1987a). *Assessing Investigations at Ages 13 and 15. Science Report for Teachers No. 9*. London, HMSO.

Department of Education and Science (1987b). *Craft, Design and Technology from 5 to 16. Curriculum Matters 9*. London, HMSO.

Department of Education and Science (1988a). *National Curriculum Task Group on Assessment and Testing: A Report*. London, HMSO.

Department of Education and Science (1988b). *Science in Schools Age 15. Review Report*. London, DES.

Department of Education and Science (1989a). *Design and Technology from 5 to 16*. London, HMSO.

Department of Education and Science (1989b). *Science in Schools Age 13. Review Report*. London, DES.

Department of Education and Science (1989c). *Science in the National Curriculum*. London, HMSO.

Doran, R.L. (1978). 'Assessing the outcomes of science laboratory activities', *Science Education*, 62, 401–9.

Douvan, E. and Adelson, J., (1966). *The Adolescent Experience*. New York, Wiley.

Dreyfus, A., Jungwirth, E. and Tamir, P. (1982). 'An approach to the assessment of teachers' concerns in the context of curriculum evaluation', *Studies in Educational Evaluation*, 8, 87–100.

Driver, R. (1983). *The Pupil as Scientist?* Milton Keynes, Open University Press.

Driver, R. and Bell, B. (1985). 'Students' thinking and the learning of science: a constructivist view', *School Science Review*, 67, 443–56.

Driver, R., Guesne, E. and Tiberghien, A. (eds) (1985). *Children's Ideas in Science*. Milton Keynes, Open University Press.

Dweck, C. *et al.* (1978). 'Sex differences in learned helplessness: the contingencies of evaluative feedback in the classroom', *Developmental Psychology*, 14, 268–76.

Eggleston, J.F., Galton, M. and Jones, M.C. (1976). *Process and Product of Science Teaching*. Schools Council Series. London, Macmillan Educational.

Eisenkraft, A.J. (1987). 'The effect of computer simulated experiments and traditional laboratory experiments on subsequent transfer task in a high school physics course', *Dissertation Abstract International*, 47, 3723.

Erickson, G. and Farkas, S. (1987). 'Prior experience: a factor which may contribute to male dominance in science'. Contributions to the Fourth GASAT Conference 2, Michigan, USA.

Fairbrother, R.W. (1988). *Assessment of Practical Work for the GCSE*. Harlow, Longman.

Fensham, P.J. (1985). 'Science for all: a reflective essay', *Journal of Curriculum Studies*, 17, 415–35.

Fensham, P.J. (ed.) (1988). *Development and Dilemmas in Science Education*. London, Falmer Press.

Fisher, D.L. and Fraser, B.J. (1982). 'Comparison of actual and preferred classroom environments as perceived by science teachers and students', *Journal of Research in Science Teaching*, 20, 55–61.

Foulds, K. and Gott, R. (1988). 'Structuring investigations in the science curriculum', *Physics Education*, 23, 347–51.

Fraser, B.J. (1978). 'Development of a test of science-related attitudes', *Science Education*, 78, 509–15.

Fraser, B.J. and Walberg, H.J. (1989). *Classroom and School Learning Environment*. London, Pergamon.

Fuhrman, M., Lunetta, V.N. and Novick, S. (1982). 'Do secondary school laboratory texts reflect the goals of the "new" science curriculum?', *Journal of Chemical Education*, 59, 563–5.

Fuhrman, M., Lunetta, V.N., Novick, S. and Tamir, P. (1978). *The Laboratory Structure and Task Analysis Inventory (LAI): A User's Handbook*. Technical Report Number 14. Iowa, Science Education Center, University of Iowa.

Fulwiler, T. (ed.) (1987). *The Journal Book*. Portsmouth, NH, Boynton/Cook.

Gagné, R.M. (1965a). *The Conditions of Learning*, 1st ed. New York, Holt, Rinehart and Winston.

Gagné, R.M. (1965b). 'The psychological basis of science – a process approach'. AAAS miscellaneous publication, 65–8. Washington DC, AAAS.

Gagné, R.M. (1970). *The Conditions of Learning*, 2nd ed. New York, Holt, Rinehart and Winston.

Gagné, R.M. and White, R.T. (1978). 'Memory structures and learning outcomes', *Review of Educational Research*, 48, 187–222.

Gamble, R., Davey, A., Gott, R., Welford, G. (1985). *Science at Age 15: Assessment of Performance Unit Science Report for Teachers, Number 5*, London, HMSO.

Ganiel, U. and Hofstein, A. (1982). 'Objective and continuous assessment of student performance in the physics laboratory', *Science Education*, 66, 581–91.

Gardner, E.D. and White, R.T. (1972). *Practical Physics: Guided Investigations for the Laboratory*. Sydney, Angus and Robertson.

Garnett, P. and O'Loughlin, M.O. (1989). 'Using practical tests to assess laboratory work in chemistry', *The Australian Science Teachers' Journal*, 35, 4, 27–37.

Gauld, C. (1986). 'Models, meters and memory', *Research in Science Education*, 16, 49–54.

Giddings, G.J. and Hofstein, A. (1980). 'Trends in the assessment of laboratory performance in high school science instruction', *Australian Science Teachers' Journal*, 62, 57–64.

Giddings G.J. and Fraser, B.J. (1989). 'Development of an instrument for assessing the psychosocial environment of science laboratory classes'. Paper presented at the Annual Conference of the Australian Science Education Research Association (ASERA), Melbourne, 30 June–3 July.

Gordon, J.E. (1976). *The New Science of Strong Materials*, 2nd edition. Harmondsworth, Penguin.

Gott, R. and Murphy, P. (1987). *Assessing Investigations at Ages 13 and 15. Assessment of Performance Unit Science Report for Teachers: 9*. London, Department of Education and Science/Welsh Office/Department of Education for Northern Ireland.

Gott, R. and Welford, G. (1987). 'The assessment of observation in science', *School Science Review*, 69, 217–27.

Gott, R., Welford, G. and Faulds, K. (1988). *The Assessment of Practical Work in Science*. Oxford, Basil Blackwell.

Gould, C.D. (1978). 'Practical work in sixth-form biology', *Journal of Biological Education*, 12, 33–8.

Grobman, H. (1970). *Developmental Curriculum Projects: Decision Points and Processes*. Itasca, Ill, Peacock Publications.

Gunstone, R.F. and Baird, J.R. (1988). 'An integrative perspective on metacognition', *Australian Journal of Reading*, 11, 238–45.

Gunstone, R.F. and Champagne, A.B. (1990). 'Promoting conceptual change in the laboratory' in Hegarty-Hazel, E. (ed.), *The Science Curriculum and the Student Laboratory*. London, Croom Helm.

Gunstone, R.F., Champagne, A.B. and Klopfer, L.E. (1981). 'Instruction for understanding: a case study', *Australian Science Teachers' Journal*, 27, 27–32.

Gunstone, R.F., Mitchell, I.J. and the Monash Children's Science Group (1988). 'Two teaching strategies for considering children's science' in *What Research Says to the Teacher: The Yearbook of the International Council of Associations of Science Education*, 1–12.

Gunstone, R.F. and White, R.T. (1981). 'Understanding gravity', *Science Education*, 65, 291–9.

Haber-Schaim, U., Cross, J.B., Dodge, J.H. and Walter, J.A. (1971). *Laboratory Guide: PSSC Physics*, 3rd ed. Lexington, MA, D.C. Heath.

Hansford, B.C. and Hattie, J.A. (1982). 'The relation between self and achievement/performance measures', *Review of Education Research*, 52, 123–42.

Hanson, N.R. (1958). *Patterns of Discovery*. Cambridge, Cambridge University Press.

Harding, J., Hildebrand, G. and Klainin, S. (1988). 'International concerns in gender and science/technology', *Educational Review*, 40, 185–93.

Harlen, W. and Jelly, S. (1989). *Developing Science in the Primary Classroom*. Edinburgh, Oliver and Boyd.

Harris, J. (ed.) (1985). *Physics Examinations and Investigations*. Revised Nuffield Advanced Science. London, Longman for the Nuffield-Chelsea Curriculum Trust.

Head, J. (1985). *The Personal Response to Science*. Cambridge, Cambridge University Press.

Heim, A.H. (1970). *AH4 Group Test of General Intelligence*. London, NFER-Nelson.

Henry, N.W. (1975). 'Objectives for laboratory work' in Gardner, P.L. (ed.), *The Structure of Science Education*. Hawthorne, Vic., Longman, pp. 61–75.

Herron, M.D. (1971). 'The nature of scientific inquiry', *School Review*, 79, 171–212.

Hildebrand, G. (1989). 'Creating a gender-inclusive science education', *Australian Science Teachers Journal*, 35, 7–16.

Hills, G.L.C. (1989). 'Students' "untutored" beliefs about natural phenomena: primitive science or commonsense?', *Science Education*, 73, 155–86.

Hobbs, E.D., Bolt, W.B., Erickson, G., Quelch, T.P. and Sieban, B.A. (1979). *British Columbia Science Assessment (1978)*, General Report, Vol. 1. Victoria, BC, Ministry of Education.

Hodgkin, R.A. (1985). *Playing and Exploring: Education through the Discovery of Order*. London, Methuen.

Hodson, D. (1982). 'Is there a scientific method?', *Education in Chemistry*, 19, 112–16.

Hodson, D. (1986). 'The nature of scientific observation', *School Science Review*, 68, 17–29.

Hodson, D. (1987). 'Social control as a factor in science curriculum change', *International Journal of Science Education*, 9, 529–40.

Hodson, D. (1988). 'Experiments in science and science teaching', *Educational Philosophy and Theory*, 2, 53–66.

Hofstein, A., Ben-Zvi, R. and Samuel, D. (1976). 'The measurement of interest in, and the attitude to laboratory work among Israeli high school chemistry students', *Science Education*, 60, 401–11.

Hofstein, A. and Lunetta, V.N. (1982). 'The role of the laboratory in science teaching: neglected aspects of research', *Review of Educational Research*, 52, 210–17.

Humrich, E. (1987). 'Girls in science: US and Japan'. Contributions to the Fourth GASAT Conference 1, Michigan, USA.

Hurd, P.D. (1969). *New Directions in Teaching Secondary School Science*. Chicago, Rand McNally.

Ingle, R.B. and Turner, A.D. (1981). 'Science curricula as cultural misfits', *European Journal of Science Education*, 3, 358–71.

Instone, I. (1988). *The Teaching of Problem-solving*. Harlow, Longman.

Jackson, K.F. (1983). *The Art of Solving Problems*. Reading, Bulmershe College.

Jenkins, E.W. (1987). 'Philosphical flaws', *Times Educational Supplement*, 2 January.

Jenkins E.W. (1989). 'Processes in science education: an historical perspective'. In Wellington, J.J. (ed.), *Skills and Processes in Science Education*. London, Routledge, pp. 21–46.

Johnson, R.T., Johnson, D.W. and Stanne, M.B. (1985). 'Effects of cooperative, competitive and individualistic goal structures on computer-assisted instruction', *Journal of Educational Psychology*, 77, 668–77.

Johnson, S. and Murphy, P. (1986). *Girls and Physics*. London, DES.

Johnstone, S.H. and Wham, A.J.B. (1982). 'The demands of practical work', *Education in Chemistry*, 19, 71–3.

Karplus, R. (1977). 'Science teaching and development of reasoning', *Journal of Research in Science Teaching*, 4, 169–75.

Kelly, A. (1981). 'Sex differences in science achievement' in Kelly, A. (ed.), *The Missing Half*. Manchester, Manchester University Press.

Kelly, A. (1987). 'Does that train set matter? Scientific hobbies and science achievement and choice'. Contributions to the Fourth GASAT Conference 1. Michigan, USA.

Kempa, R.F. and Palmer, C.R. (1974). 'The effectiveness of video-tape recorded demonstrations in the learning of manipulative skills in practical chemistry', *British Journal of Educational Technology*, 1, 62–70.

Kempa, R.F. and Ward, J.E. (1975). 'The effect of different mode of task orientation on observational attainment of practical work', *Journal of Research in Science Teaching*, 12, 69–76.

Khor, S.Y. (1987). *A Study of Academic Self-Concept and its Relation to Academic Achievement in Two Samples of Singaporean Pupils*. Singapore, Institute of Education.

Kimbell, R. (1988). 'The Assessment of Performance Unit project in Design and Technology'. In *DATER 88*. Proceedings of the First National Conference in Design and Technology Education Research and Curriculum Development, Loughborough University.

King, K. (1986). 'Mapping the environment of science in India', *Studies in Science Education*, 13, 53–69.

Kinnear, J.F. (1982). 'Computer simulation and concept development in students of genetics', *Research in Science Education*, 12, 89–96.

Klopfer, L.E. (1971). 'Evaluation of learning in science' in Bloom, B.S., Hastings, J.T. and Maddaus, G.F. (eds), *Handbook on Summative and Formative Evaluation of Student Learning*. New York, McGraw-Hill.

Knamiller, G.W. (1984). 'The struggle for relevance in science education in developing countries', *Studies in Science Education*, 11, 60–77.

Knamiller, G.W. (1988). 'Linking traditional science and technology with school science in Malawi. Mimeo, University of Leeds.

Koroma, A.K. (1975). The attitudes of secondary school science teachers in Sierra Leone towards modern methods of science teaching. M.Ed. dissertation, University of Birmingham.

Krajcik, J.S., Simmons P.E. and Lunetta, V.N. (1986). 'Improving research on computers in science learning', *Journal of Research in Science Teaching*, 23, 465–70.

Kuehl, E. (1984). 'An evaluation of high school physics laboratory manuals', *The Physics Teacher*, 22, 222–34.

Kuhn, T.S. (1977). 'Second thoughts on paradigms' in *The Essential Tension*. Chicago, University of Chicago Press, pp. 293–319.

Lakatos, I. (1970). 'Falsification and the methodology of scientific research programmes' in Lakatos, I. and Musgrave, A. (eds), *Criticism and the Growth of Knowledge*. Cambridge, Cambridge University Press.

Law, J. and Lodge, P. (1984). *Science for Social Scientists*. London, Macmillan.

Lawson, A.E. (1975). 'Developing formal thought through biology teaching', *American Biology Teacher*, 37, 411–19, 429.

Lawson, A.E., Abraham, M.R. and Renner, J.W. (1989). *A Theory of Instruction*, NARST Monograph No. 1.

Layton, D. (1973). *Science for the People*. London, Allen & Unwin.

Layton, D., Davey, A. and Jenkins, E. (1986). 'Science for Specific Social Purposes (SSSP): perspectives on adult scientific literacy', *Studies in Science Education*, 13, 27–52.

Lee, M.N.N. (1982). 'Science education in Malaysian schools: some problems and suggestions, *Dan Pendikan*, 4, 58–63.

Lock, R. and Wheatley, T. (1989). 'Recording process skills and criterion assessments – student systems', *School Science Review*, 71, 255.

Lunetta, V.N. and Blick, D.J. (1973). 'Evaluation of a series of computer based dialogues in introductory physics, *Journal of the Association for Educational Data Systems*, 33, 42.

Lunetta, V.N. and Hofstein, A. (1981). 'Simulations in science education', *Science Education*, 65, July, 243–53.

Lunetta, V.N. and Peters, H.J. (1985). 'Simulations in education', *Curriculum Review*, 24, March/April, 30–4.

Lunetta, V.N. and Tamir, P. (1979). 'Matching lab activities with teaching goals', *The Science Teacher*, 46, May, 22–4.

Lyth, M. (ed.) (1986). *Nuffield 11–13 Science. How Scientists Work, How Science is Used. With Teachers' Guides*. London, Longman.

Mackenzie, A.A. (1979). 'Geographical Fieldwork and Long Term Memory Structures'. Unpublished master's thesis, Melbourne, Monash University.

Mackenzie, A.A. and White, R.T. (1982). 'Fieldwork in geography and long term memory structures', *American Educational Research Journal*, 19, 623–32.

McComas, W.F. (1989). 'The application of scientific knowledge: the results of the 1987–88 Chautauqua workshops', *Chautauqua Notes*, 4, 1–2.

McComas, W.F. and Yager, R.E. (1988). 'Preference and Understanding Instrument' in *The Iowa Assessment Package for Evaluation in Five Domains of Science Education*. Iowa City, The University of Iowa, Science Education Center, pp. 71–89.

Medawar, P.B. (1969). *Induction and Intuition in Scientific Thought*. London, Methuen.

Millar, R. (1987). 'Towards a role for experimentation in the science teaching laboratory', *Studies in Science Education*, 14, 109–18.

Millar, R. (1988). 'What is scientific method and can it be taught?' in Wellington, J.J. (ed.), *Skills and Processes in Science Education*. London, Routledge.

Millar, R. and Driver, R. (1987). 'Beyond processes', *Studies in Science Education*, 14, 33–62.

Millar, J.D. (1988). 'Educators rework the way schools teach science', *Christian Science Monitor*, 80, 21–2.

Millar, J.D., Suchner, R.W. and Voelker, A.M. (1980). *Citizenship in an Age of Science: Changing Attitudes among Young Adults*. New York, Pergamon Press.

Mitchell, I.J. and Baird, J.R. (eds) (in press). *Teaching for Intellectual Independence*. Faculty of Education, Monash University.

Moreira, M.A. (1980). 'A non-traditional approach to the evaluation of laboratory instruction in general physics courses', *European Journal of Science Education*, 2, 441–8.

Moscovici, S. (1976). *Social Influence and Social Change*. European Monographs in Social Psychology. London, Academic Press.

Mulkay, M. (1979). *Science and the Sociology of Knowledge*. London, George Allen & Unwin.

Murphy, P. (1985). Personal communication, 3 July.

Murphy, P. and Gott, R. (1984). *Science Assessment Framework Age 13 and 15*, Science Report for Teachers: 2. ASE

Myers, L.H. (1988). 'Analysis of student outcomes in ninth grade physical science taught with a Science Department Science/Technology/Society focus versus one taught with a textbook orientation'. Unpublished doctoral dissertation, Iowa City, University of Iowa.

Nachmias, R. and Linn, M.C. (1987). 'Evaluation of science laboratory data: the role of computer presented information', *Journal of Research in Science Teaching*, 24, 491–506.

National Assessment of Educational Progress (1978a). *The Third Assessment of Science*, *1976–77*. Denver, CO, NAEP.

National Assessment of Educational Progress (1978b). *Science Achievement in the Schools*. A summary of results from the 1976–77 National Assessment of Science, Washington Education Commission of The States, Washington DC.

National Science Teachers Association (1990). *The NSTA Position Statement on Science/Technology/Society (STS)*. Washington, DC, NSTA.

Norris, S.P. (1983). 'The philosophical basis of observation in science and science education', *Journal of Research in Science Teaching*, 22, 817–33.

Norris, S.P. (1984). 'Defining observational competence', *Science Education*, 68, 129–42.

Novak, J.D. (1978). 'An alternative to Piagetian psychology for science and mathematics education', *Studies in Science Education*, 5, 1–30.

Novak, J.D. and Gowin, D.B. (1984). *Learning How to Learn*. Cambridge, Cambridge University Press.

OCEA (1987). *OCEA Science: Teachers' Guide*. Oxford, Oxford University Press.

Oliver, P.M. and Roberts, I.F. (1974). 'Filmed sequences and multiple choice tests', *Education in Chemistry*, 11, 132–3.

Omerod, M.B. (1981). 'Factors differentially affecting the science subject preferences, choices and attitudes of girls and boys' in Kelly, A. (ed.), *The Missing Half*. Manchester, Manchester University Press.

Osborne, R. (1976). 'Using student attitudes to modify instruction in physics', *Journal of Research in Science Teaching*, 13, 525–31.

Osborne, R.J. and Freyberg, P. (eds) (1985). *Learning in Science: The Implications of Children's Science*. Auckland, Heinemann.

Osborne, R.J. and Wittrock, M.C. (1985). 'The generative learning model and its implications for science education', *Studies in Science Education*, 12, 59–87.

Page, R. (1983). 'The Schools Council's 'Modular Courses in Technology' Project', in *The Stanley Link in Craft Design and Technology*. Stanley Tools.

Paris, S.G., Cross, D.R. and Lipson, M.Y. (1984). 'Informed strategies for learning: a program to improve children's reading awareness and comprehension', *Journal of Educational Psychology*, 76, 1239–52.

Pella, M.D. (1961). 'The laboratory and science teaching', *The Science Teacher*, 28, 20–31.

Penick, J.E., Shymansky, J.A., Filkins, K.M. and Kyle, W.C. (1976). *Science Laboratory Interaction Category (SLIC) – Student*. Iowa City, Science Education Center.

Pickering, M. (1980). 'Are lab courses a waste of time?', *Chronicle of Higher Education*, 19, February.

Polanyi, M. (1958). *Personal Knowledge*. London, Routledge and Kegan Paul.

Popper, K.R. (1959). *The Logic of Scientific Discovery*. London, Hutchinson.

Pouler, C.A. and Wright, E.L. (1980). 'An analysis of the influence of reinforcement and knowledge of criteria on the ability of students to generate hypotheses', *Journal of Research in Science Teaching*, 17, 31–7.

Randall, G.J. (1987). 'Gender differences in pupil–teacher interactions in workshops and laboratories' in Weiner, G. and Arnot, M. (eds), *Gender under Scrutiny*. Milton Keynes, Open University Press.

Ravetz, J.R. (1971). *Scientific Knowledge and its Social Problems*. Oxford, Oxford University Press.

Rennie, L.J. (1987). 'Out of school science: are gender differences related to subsequent attitudes and achievement in science?'. Contributions to the Fourth GASAT Conference 2. Michigan, USA.

Richmond, P.E. (1978). 'Who needs laboratories?' *Physics Education*, 14, 349–50.

Robinson, T.J. (1969). 'Evaluating laboratory work in high school biology', *The American Biology Teacher*, 34, 226–9.

Rowell, J.A. (1984). 'Towards controlling variables: a theoretical appraisal and a teachable result', *European Journal of Science Education*, 6, 115–30.

Rutherford, F.J., Holton, G. and Watson, F.G. (1970). *The Project Physics Course Handbook*. New York, Holt, Rinehart and Winston.

Sargent, J.G. (1989). Explorations in science horizons: a case study of changing teaching styles. PhD. thesis, King's College, University of London.

Schibeci, R.A. (1984). 'Attitudes to science: an update', *Studies in Science Education*, 11, 26–59.

Schools Council (1970). *The Next Two Years*. Schools Council Project Technology. London, Schools Council.

Schools Council (1975). *Education through Design and Craft*. Schools Council Design and Craft Education Project. London, Schools Council.

School Technology Forum (1974). 'The essence of school technology' in *Technology Interface, School and Teacher Training*. London, Standing Conference on School Science and Technology.

Schwab, J.J. (1960). 'Enquiry – the science teacher and the educator', *The Science Teacher*, 27, 6–11.

Schwab, J.J. (1962). 'The teaching of science as enquiry' in Schwab, J.J. and Brandwein, P.F. (eds), *The Teaching of Science*. Cambridge, MA, Harvard University Press.

Screen, P. (ed.) (1986a). *Warwick Process Science*. Southampton, Ashford Press.

Screen, P. (1986b). 'The Warwick Process Science Project', *School Science Review*, 68, 12–16.

SEC (1986). *Craft Design and Technology GCSE. A Guide for Teachers*. Milton Keynes, Open University.

Selmes, C., Ashton, B.G., Meredith, H.M. and Newal, A. (1969). 'Attitudes to science and scientists', *School Science Review*, 51, 7–22.

Shipstone, D.M. and Gunstone, R.F. (1985). 'Teaching children to discriminate between current and energy' in Duit, R., Jung, W. and von Rhöneck, C. (eds), *Aspects of Understanding Electricity*. Kiel, W. Germany, Schmidt und Klaunig, pp. 187–98.

Shulman, L. and Tamir, P. (1973). 'Research on teaching in the natural sciences' in Travers, R.M.W. (ed.), *Second Handbook of Research on Teaching*. Chicago, Rand McNally, pp. 1098–1148.

Shymansky, J.A., Penick, J.E., Kelsey, L.J. and Foster, G.W. (1976). *Science Laboratory Interaction Category (SLIC) – Teacher*. Iowa City, Science Education Center.

Sim, W.K. (1977). 'Evaluation of integrated science teaching in Malaysia' in *New Trends in the Teaching of Integrated Science: Volume 4*. Paris, Unesco, pp. 173–85.

Simmons, P.E. and Lunetta, V.N. (1987). 'CATLAB – a learning cycle approach', *The American Biology Teacher*, 49, February, 107–9.

Simpson, R. (1972). *Simple Physics Apparatus.* Department of Education, Hong Kong University.

Skurnik, L.S. and Jeffs, P.M. (1970). *Science Attitude Questionnaire.* Slough, NFER.

Slavin, R.E. (1980). 'Cooperative learning', *Review of Educational Research*, 50, 241–71.

Smith, J.P. (1971). 'The development of a classroom observation instrument relevant to the earth science curriculum project', *Journal of Research in Science Teaching*, 89, 231–5.

Solomon, J. (1980). *Teaching Children in the Laboratory.* London, Croom Helm.

Solomon, J. (1988). 'Learning through experiment', *Studies in Science Education*, 15, 103–8.

Solomon, J. (1989). 'A study of behaviour in the teaching laboratory', *International Journal of Science Education*, 11, 317–26.

Stevens, P. (1978). 'On the Nuffield philosophy of science', *Journal of Philosophy of Science*, 12, 99–111.

Swift, D.G. (1983). *Physics for Rural Development: A Sourcebook for Teachers and Extension Workers in Developing Countries.* Chichester, John Wiley.

Tamir, P. (1972). 'The practical mode – a distinct mode of performance in biology', *Journal of Biological Education*, 6, 175–82.

Tamir, P. (1974). 'An inquiry-oriented laboratory examination', *Journal of Educational Measurement*, 11, 23–5.

Tamir P. (1977). 'How are laboratories used?', *Journal of Research in Science Teaching*, 14, 311–16.

Tamir, P. (1989a). Personal communication. 12 October 1989.

Tamir, P. (1989b). 'Gender differences in science education as revealed by the Second International Science Study'. Contributions to the Fifth GASAT Conference. Haifa, Israel.

Tamir, P. (1989c). 'Teaching effectively in the laboratory', *Science Education*, 73, 59–69.

Tamir, P. and Lunetta, V.N. (1978). 'An analysis of laboratory activities', in the BSCS yellow version, *The American Biology Teacher*, 40, 353–7.

Tamir, P. and Lunetta, V.N. (1981). 'Inquiry related tasks in high school science laboratory handbooks', *Science Education*, 65, 477–84.

Tasker, R. (1981). 'Children's views and classroom experiences', *Australian Science Teachers' Journal*, 27, 33–7.

Thompson, J.J. (ed.) (1975). *Practical Work in Sixth Form Science.* Oxford, Department of Educational Studies, University of Oxford.

Tobin, K. (1984). 'Student engagement in science learning tasks', *European Journal of Science Education*, 6, 339–47.

Tobin, K., Pike, G. and Lacey, T. (1984). 'Strategy analysis procedures for improving the quality of activity oriented science teaching', *European Journal of Science Education*, 6, 79–89.

Toh, K.A. and Woolnough, B.E. (1990). 'Assessing, through reporting, the outcomes of scientific investigations', *Educational Research*, 32, 42–8.

Torrance, P.E. (1966). *Torrance Test of Creative Thinking.* Princeton, NJ, Personnel Press Inc.

Unesco (1973). *New Unesco Sourcebook for Science Teachers.* Paris, Unesco/Heinemann.

Varghese, T. (1988). 'Problems in teaching chemistry in Zambian secondary schools' in Thulstrup, E.W. (ed.), *Teaching Chemistry at Low Cost.* Paris, Unesco.

Walberg, H.J. (1970). 'A model for research on instruction', *School Review*, 78, 185–99.

Wallace, J. (1986). Social interaction within second year groups. Unpublished MSc thesis, University of Oxford.

Wellington, J.J. (1981). 'What's supposed to happen, Sir?', *School Science Review*, 63, 167–73.

West, L.H.T. and Fensham, P.J. (1974). 'Prior knowledge and the learning of science: a review of Ausubel's theory of this process', *Studies in Science Education*, 1, 61–81.

West, L.H.T. and Pines, A.L. (eds) (1985). *Cognitive Structure and Conceptual Change*. Orlando, FL, Academic Press.

White, R.T. (1982). 'Memory for personal events', *Human Learning*, 1, 171–83.

White, R.T. (1988). *Learning Science*. Oxford, Blackwell.

White, R.T. (1989). 'Recall of autobiographical events', *Applied Cognitive Psychology*, 3, 127–35.

White, R.T. and Gunstone, R.F. (1980). 'Converting memory protocols to scores on several dimensions' in *Annual Conference Papers*, Australian Association for Research in Education, pp. 486–93.

Wightman, T., Green, P. and Scott, P. (1986). *Case Studies on the Particulate Nature of Matter*. Leeds, University of Leeds.

Williams, I.W. (1979). 'The implementation of curricula adapted from Scottish Integrated Science' in P. Tamir *et al* (eds), *Curriculum Implementation and its Relationship to Curriculum Development in Science*. Jerusalem, Israel Science Teaching Centre, pp. 295–9.

Willmott, A.S., Bennetts, J. and Fairbrother, R.W. (1987). 'The moderation of teacher assessment', *Education in Science*, 121, January.

Wilson, B. (1983). 'Teaching styles and teacher education'. Lead paper for the international symposium on the Cultural Implications of Science Education, Zaria, Nigeria.

Woolnough, B.E. (1986). 'The assessment of practical work'. *Physics Education*, 21, 195–6.

Woolnough, B.E. (1988a). *Physics Teaching in Schools 1960–1985*. Lewes, The Falmer Press.

Woolnough, B.E. (1988b). 'Reductio ad absurdum?', *Physics Education*, 23, 1–2.

Woolnough, B.E. (1989a). 'Towards a holistic view of processes in science education'. In Wellington, J.J. (ed.), *Skills and Processes in Science Education*. London, Routledge, pp. 115–34.

Woolnough, B.E. (1989b). 'Risk of drowning by numbers', *The Independent*, 2 November.

Woolnough, B.E. and Allsop, T. (1985). *Practical Work in Science*. Cambridge, Cambridge University Press.

Woolnough, B.E. and Toh, K.A. (1990). 'Alternative approaches to assessment of practical work in science', *School Science Review*, 71, pp. 127–31.

World Bank (1988). *Education in Sub-Saharan Africa*. Washington, DC, World Bank.

Wray, J. *et al.* (eds) (1987). *Science in Process*. Ten units and Teachers' Guide. London, Heinemann.

Yager, R.E. (in press). 'Development of student creative skills: a quest for a successful science education', *Creativity Research Journal*.

Yager, R.E. and Bonnstetter, R.J. (1984). 'Student perceptions of science teachers, classes, and course content', *School Science and Mathematics*, 84, 406–14.

Yager, R.E. and Penick, J.E. (1986). 'Perceptions of four age groups toward science classes, teachers, and the value of science', *Science Education*, 70, 355–63.

Yager, R.E. and Wick, J.W. (1966). 'Some aspects of the student's attitude in science courses', *School Science and Mathematics*, 66, 269–73.

Young, B.L. (1979). *Teaching Primary Science*. London, Longman.

Zambia Education Reform Document (1977). Lusaka, Government Printer.

Index

affective factors, 7, 14, 90, 100, 187
aims of practical work, 4, 5, 14–18, 31,
    32, 43, 44, 51, 52, 130, 181, 182
alienation, 117–21
analogues, 128
application, 26, 27
assessment
    atomistic and holistic, 140, 145–7, 154,
        155, 160, 172, 184, 185
    in context, 156–60
    continuous, 142, 171
    effects of, 8, 9, 35, 184
    objectives, 163, 164, 166–7
    techniques, 8, 9, 140, 168–72
Assessment of Performance Unit (APU)
    design and technology, 141–4
    science, 91, 113–15, 158, 159
atomistic, and holistic approaches, 5, 154,
    155, 187
attainment targets, 161
Attainment Target 1, exploration of
    science, 65, 66, 161
attitudes
    to laboratory work , 172–4, 187
    to school, 92
    to science, 28, 92
    to self, 92, 117, 118

balanced curriculum, 15
behaviour of students in laboratory, 13,
    101–11 passim

capability, 144, 145
classifying, 48, 49
community-based investigations, 36
computer/simulation assessment, 128
computers in the laboratory, 127–30
constructivist model of learning, 56, 67,
    71–6
coursework assessment, 155–65 passim
craft activity, science as, 6, 46
creativity skills, 27, 28
cultural factors, 34, 35, 67
curriculum materials, 18–20, 61–3

danger, of experiments, 126, 135
demonstrations, 69
dependent variables, 58
design, 140–44
differentiation by outcomes, 163
discovery, 14
discussion, 102, 103, 108, 109
doing science, being good at, 5

emotional support, 102–9
environmental approaches, 24, 36
environment of laboratory, 172–4
episodes, 78–84
ethnomethodology, for laboratory
    analysis, 101
evaluation, 174–7
examination, 155
experience, 116–18

experiences, 75, 187
experiential learning, 5
explicit instruction, 95–100
explicit knowledge, 90, 91
explorations, 186, 187
explore with, 110

gender differences, 99, 112–22, 150
Graded Assessments in Science Project
    (GASP), 158
group learning, 107, 108
group relationships, 105–8

holistic, assessment, 185
    model of learning, 161, 185–8
hypothetico-deductive view of science, 46

improvised apparatus, 37, 38
independent variables, 58
individual, and the group, 108, 109
inductive reasoning, 46
information processing, 80
instructional packages, 27, 34, 68–70,
    95–100
inquiry, 14
inquiry laboratories, 16–18
inquiry tactics, 51
internal modelling, 109
investigations, 5, 16, 89–100 *passim*
iterative approach, 143, 159

knowledge, types of, 79, 80
knowledge refinement model, 56, 57
knowledge transmission model, 55, 56

Laboratory Analysis Inventory (LAI), 18,
    176
laboratory manuals, 19, 174–6
laboratory, of schools and of scientists, 16
learning
    cycle, 17
    styles, 75–6, 85
    theories, 79, 90
learning, through instruction, 27, 34,
    68–70, 95–100
linking, 76, 82–5
listening to student talk, 102, 103
local equipment production, 38

metacognition, 76
methods of science, 5, 46, 58, 59, 185–8
misconceptions, 14
modelling, 110, 129, 142

models, 129
moderation of teacher assessments, 165
motivation, 63–5, 183

National Curriculum, 65–6, 161
negotiating perception, 103
Nuffield schemes, 55, 56, 163

observation, and inference, 72
    nature of, 47, 48, 71–4
    and personal theories, 72
    student understanding of, 73
    theory-laden, 47, 48, 71, 72, 108
optimising, 149
ownership, 105, 106, 182
Oxfordshire Certificate of Educational
    Achievement (OCEA), 160

personal analogues, 67
playing, practice and exploring, 110, 186
Polanyi, 6, 91, 110
Popper, 46
practical examinations, 155, 168–70
practical skills, 28, 45, 155, 168
preconception
    of science, 67–74 *passim*
    of teaching and learning, 75, 76
predict/observe/explain (POE), 69–77
    passim
prior knowledge, 89, 90
private, and public behaviour, 103–9
    passim
problem perception, 119, 120
problem-solving, 14, 25, 58 , 82, 130, 141,
    158
problem-solving model (APU), 8, 159, 182
procedural understanding, 58, 59, 145, 146
process approach, 44–6, 57, 58, 141
processes, 62
process skills, 27–9, 45
professional judgement, 165, 166
professional training, 38–40, 165
progression, 8, 63, 64, 97, 160–2, 183, 184
propositions, 79
psychology of student grouping, 106, 107

real-world problems, 24
reductionism, 185
relationship between student factors, 99
relationship between theory and practical,
    6, 14, 15, 47–50, 65, 66, 71–5, 79, 80,
    95–8, 142, 143, 182, 183
reliability, 162–5

report sheets, 92–3
resources, 33, 34, 37
restructuring theories, 68–77 *passim*

Science Education Project for Africa
    (SEPA), 39
Science Equipment Production Unit,
    Kenya (SEPU), 38
science kits, 38
Science Laboratory Environment
    Inventory (SLEI), 175
science/technology literacy, 24, 25
Science/Technology/Society (STS), 21–4,
    26
scientific activity, nature of, 5, 6, 16, 46
scientifically literate students, 24, 25
secondary sources, 128
self-concept, 90
self-confidence, 7, 187
self-image, 104, 105
simulation, 125–7, 170, 171
social control, 54, 55
social development, 105–8
step-up model of learning, 5
stratification of the curriculum, 59–61
subject disciplines, 19

tacit knowledge, 6, 90–100 *passim*, 187
talk, during practical work, 102–8 *passim*
tasks, 61–4, 147–50
teacher assessment, 153
teacher demonstrations, 69
teachers, 20, 176, 177
Techniques for Assessing Practical Skills
    (TAPS), 45
technology, 24, 138–50
transferability, 7, 45, 156–60, 183
transferable skills, 45
transmission of knowledge, 14–15
types of practical work
    demonstrations, 69
    episodes, 78–81
    experiences, 44, 75, 187
    investigations, 44
    tasks, 61–4, 147–50

validity, 162–5
verification exercises, 15, 73

whole, greater than sum of the parts, 185
within-group tutoring, 109

ZimSci, 34, 38, 39